D1526775

EMI/EMC COMPUTATIONAL MODELING HANDBOOK

EMI/EMC COMPUTATIONAL MODELING HANDBOOK

by

Bruce Archambeault
IBM Corporation.

and

Omar M. Ramahi
Colin Brench
Digital Equipment Corporation

KLUWER ACADEMIC PUBLISHERS
Boston / Dordrecht / London

Distributors for North, Central and South America:
Kluwer Academic Publishers
101 Philip Drive
Assinippi Park
Norwell, Massachusetts 02061 USA

Distributors for all other countries:
Kluwer Academic Publishers
Distribution Centre
Post Office Box 322
3300 AH Dordrecht, THE NETHERLANDS

Library of Congress Cataloging-in-Publication Data
Archambeault, Bruce.
EMI/EMC computational modeling handbook / by Bruce Archambeault, Omar Ramahi, Colin Brench.
p. cm.
Includes bibliographical references and index.
ISBN 0-412-12541-2 (hb : alk. paper)
1. Electromagnetic interference--Mathematical models.
2. Electromagnetic compatibility--Mathematical models. I. Ramahi, Omar. II. Brench, Colin. III. Title.
TK7867.2.A73 1997
621.382'24'015194--dc21 97-43129
CIP

Printed on acid-free paper.

Printed in the United States of America

Dedication

I wish to dedicate this book to the memory of my parents, Wilfred and Margueritte Archambeault. Without their support during my early years, I doubt I could have achieved all that I have achieved. I only wish they had seen this book.

—Bruce Archambeault

To my parents, who in their own way made this work possible.

—Omar M. Ramahi

To Bronwyn

—Colin Brench

Contents

Preface

The application of computational electromagnetics to EMI/EMC engineering is an emerging technology. With the advancement in electronics, EMI/EMC issues have greatly increased in complexity. As a result, it is no longer possible to rely exclusively on traditional techniques and expect cost-effective solutions. This book introduces computational electromagnetics to EMI/EMC engineering.

A number of books are available on electromagnetic theory, others on computational techniques, and still others on EMI/EMC engineering. This book combines the essential elements of these three fields, rather than presenting a comprehensive treatment of each. It is intended to provide an understanding for those interested in incorporating modeling techniques in their work. Modeling is not an exact science, and therefore no book on modeling can ever be comprehensive. However, it is hoped that this book, through the experience of the authors, presents a philosophy of modeling that can be applied for a wide range of applications. To the best of our knowledge, no book has yet been published on the use of computational electromagnetic tools as applied to EMI/EMC problems. This book is intended to fill this void.

This book will serve many different levels of readers. It will serve as a basic introduction to modeling as applied to EMI/EMC problems for the engineer interested in getting started, and it will help the person already using modeling as a tool to become more effective in using different modeling techniques. It will also be useful for the engineer who is familiar with computational techniques and wishes to apply them to EMI/EMC applications. This book can also be used as a text to help students of electromagnetic theory and application better understand real-world challenges facing engineers.

Chapter 1 sets the overall tone of this book by providing an overview of the source of EMI/EMC problems and what is possible with the current state of the art in modeling. Complete systems with all the individual components cannot be included in the numerical models; however, useful information can be obtained when numerical models are properly constructed. Numerical modeling does not replace the need for the EMI/EMC engineer, but provides an additional tool to the engineer to help achieve successful designs more rapidly. Chapter 2 provides a brief introduction to electromagnetic theory, providing the pertinent equations which form the foundations for the numerical modeling technique discussed in the following chapters. Special emphasis is placed on the time-varying and frequency domain forms of Maxwell's equations.

The next three chapters present three popular numerical modeling tools. Chapter 3 gives a thorough description of the Finite-Difference Time-Domain (FDTD) method. FDTD is a relatively new method in EMI/EMC modeling and therefore, more details on this technique are provided than the other methods. Chapter 4 covers the Method of Moments (MoM) technique. While MoM has been used for years in EMI/EMC applications, available literature on this technique are focused on different applications. Chapter 5 presents the third modeling technique, the Finite Element Method (FEM). For both the MoM and the FEM, sufficient details are included to allow the reader to gain an understanding of these techniques. Readers need only a basic understanding of linear algebra and vector calculus to be able to follow these chapters.

Chapter 6 gets to the heart of the EMI/EMC numerical modeling problems. A discussion on the usefulness of the various modeling techniques for different types of modeling problems is presented. Discussions on selecting time-domain or frequency-domain, and quasi-static or full-wave techniques are provided. The reader is presented with an understanding of how to model the source of the EMI/EMC energy, and what goals are appropriate for EMI/EMC models. Chapter 7 provides a step-by-step approach to creating EMI/EMC numerical models for each of the various modeling techniques. A number of examples are given using a step-by-step approach.

Chapter 8 is a discussion of special topics in EMI/EMC numerical modeling. A number of more advanced techniques and applications are given. Chapter 9 discusses model validation. Models can be validated at a number of different levels, from modeling technique validation, to specific computer code validation, to individual model validation. Techniques to help the reader validate their models are presented. Chapter 10 provides a number of standard EMI/EMC modeling problems. Standard models are provided which can be used as a starting point for software evaluation. These standard models can also be used for benchmarking.

The authors gratefully acknowledge the editorial and technical review by Bronwyn Brench, whose EMC engineering background and editorial skills are a valuable combination that enhanced the quality of this book. Gratitude is also due to Dr. Greg Hiltz, of Isotec Corporation, in Ottowa, Canada, for his assistance in creating special software applications used for a number of the examples in this book, and for his technical review of the book. Finally, this book would not have been possible without the support of our dear wives; Susan Archambeault, Idrisa Pandit, and Bronwyn Brench. Their patience during the authors mood swings of desperation, exuberance, and despair is greatly appreciated. Without their continuing support and understanding, this book would have never been completed.

Chapter 1

Introduction

1.0 Introduction to EMI/EMC

Electromagnetic Interference and Electromagnetic Compatibility (EMI/EMC) first became a concern during the 1940s and 1950s, mostly as motor noise that was conducted over power lines and into sensitive equipment. During this period, and through the 1960s, EMI/EMC was primarily of interest to the military to ensure electromagnetic compatibility. In a few notable accidents, radar emissions caused inadvertent weapons release, or EMI caused navigation systems failure, and so military EMI/EMC was concerned chiefly with electromagnetic compatibility, especially within a weapons system, such as a plane or ship.

With the computer proliferation during the 1970s and 1980s, interference from computing devices became a significant problem to broadcast television and radio reception, as well as emergency services radio reception. The government decided to regulate electromagnetic emissions from products in this industry. The FCC created a set of rules to govern the amount of emissions from any type of computing device, and how those emissions were to be measured. Similarly, European and other governments began to control emissions from electronic and computing devices. However, during this time, EMI/EMC control was limited to computers, peripherals, and computer communications products.

During the 1990s, the concern over EMI/EMC has been found to broaden dramatically; in fact, many countries have instituted import controls requiring that EMI/EMC regulations be met before products can be imported into that country. The overall compatibility of all devices and equipment must coexist harmoniously in the overall electromagnetic environment. Emissions, susceptibility to emissions from other equipment, susceptibility to electrostatic discharge—all from either radiated or conducted media—are controlled. No longer is this control limited to only computers, but now any product that may potentially radiate EMI, or that could be susceptible to other emissions, must be carefully tested. Products with no previous need for EMI/EMC control must now comply with the regulations, including dishwashers, videocassette recorders (VCRs), industrial equipment, and most electronic equipment.

While commercial products have come under tighter control for the military has not relaxed its EMI/EMC requirements. In fact, because of the higher degree of automation and faster processing speeds, military EMI/EMC control has become a significant part of all military programs.

EMI/EMC design means different things to different people. The standards for commercial applications, such as VCRs, personal computers, and televisions are fairly loose compared to military/TEMPEST[1] standards. However, they are still difficult to meet—as a result of the relaxed nature of these commercial standards, designers are constantly caught between lowering emissions and susceptibility while reducing costs. The tradeoffs between EMI/EMC design features are clear, but whether one or another individual EMI/EMC component is required is very unclear. Traditionally, EMI/EMC engineers have used experi-

[1]TEMPEST is the U.S. government code name for the project that controls data-related radio frequency (RF) emissions from equipment processing classified information.

ence, as well as equations and graphs from handbooks, frequently taken out of context, to help during the product design phase. Very little high-quality EMI/EMC engineering-level training is available at universities, or at any institutions, and most engineers working in this area find these present methods somewhat inadequate.

Military, space, and other government applications must control the emissions of electronics, for security, weapon systems functionality, or proper communications, most often to a level far below the commercial emissions/susceptibility level. This increased control requires additional EMI/EMC design features, and greater expense, just when these applications are being forced to reduce costs.

EMI/EMC problems are caused by changes in current with respect to time on conductors within the equipment, known as *di/dt* noise. This current change causes electromagnetic emissions. Alternatively, external electromagnetic energy can induce *di/dt* noise in circuits, causing false logic switching and improper operation of devices. Most high-speed fast rise time signals cause EMI/EMC problems. These problems are enhanced through the wires and cables attached to the product, creating more efficient antennas at lower frequencies. The normal solution is to use metal shielding, to filter all data/power lines, and to provide significant on-board filtering of signal lines and power planes. The real question is "how much is enough?" and "how much is too much?"

The subject of EMI modeling is beginning to appear in the technical literature with increasing frequency. Most articles identify some new feature or special model that may or may not apply to the general EMI/EMC engineer responsible for product development. Little information is available to the potential user of EMI/EMC modeling tools without requiring reading textbooks and technical papers containing lots of heavy mathematics and advanced electromagnetic theory.

The current state of the art in EMI/EMC modeling, however, does not require an engineer to have advanced training in electromagnetics or numerical modeling techniques before accurate simulations can be performed and meaningful results obtained. Modeling of EMI/EMC problems can truly help the typical engineer but, like any tool, before modeling can be used effectively, the basics must be understood.

This book will serve many different levels of readers. It will serve as a basic introduction to modeling as applied to EMI/EMC problems for the engineer interested in getting started, and it will help the person already using modeling as a tool to become more effective in using different modeling techniques. It will also be useful for the engineer familiar with computational techniques who wishes to use them in EMI/EMC applications. A description of the most common different EMI/EMC modeling techniques will be provided along with their relative strengths and weaknesses. Examples of each of these techniques will be given that typically appear as real-world problems to the EMI/EMC engineer. Standard problems will be presented to allow interested users to evaluate vendor software before purchase to ensure that it can simulate the types of problems EMI engineers experience.

1.1 Why Is EMI/EMC Modeling Important?

The main reason to use EMI/EMC modeling as one of the tools in the EMI/EMC engineer's tool box is to reduce the cost of the product.[2] Without modeling, engineers must rely on handbooks, equations, and graphs, all of which have limited applicability, as well as their own rules of thumb, developed through experience. These guidelines are usually based on assumptions that frequently do not exist in the problem at hand. Some guidelines are better, in that they attempt to correct for the inappropriate assumptions, but even these can have limitations due to accuracy in all but the most carefully controlled circumstances. Proper use of modeling tools allow engineers to use a full-wave electromagnetic solution, rather than one or more simplifications, to predict the effect in the specific product of concern.

Given the limitations that these guidelines have in most real-world problems, engineers are faced with either a conservative or nonconservative design. The conservative design will ensure that the product will meet the appropriate regulatory limits the first time. This can be ensured only by overdesign of the EMI/EMC features. This overdesign will usually meet the appropriate limits, but extra cost is added to the product. The nonconservative design will take some reasonable chances to reduce the amount of EMI/EMC features required. Depending on the engineer's experience and training, the product may or may not meet the regulatory limit. If the product doesn't meet the limit, a panic

[2]The cost of the product can be measured in both the development costs and the time-to-market costs.

redesign is required, most often resulting in product ship delays and extra cost due to the band-aid nature of such "fixes."

Another realistic benefit of the use of EMI/EMC modeling is credibility. Often the product design team consists of a number of different engineering disciplines: electrical, mechanical, thermal, and EMI/EMC. Computer Aided Design (CAD) simulation tools are commonly used in other engineering disciplines. These tools provide significant credibility to the engineer's claim for whatever design features they recommend to be included for a successful EMI/EMC design. These features, such as larger air vent openings, are often in direct conflict with the EMI/EMC engineer's design direction; however, since the EMI/EMC engineer has no simulation to rely on, their recommendations are often ignored. EMI/EMC modeling tools can provide the design team with reliable numerical results, taking the guesswork out of the design and providing the EMI/EMC engineer with the credibility to get their design recommendations seriously considered by the team.

1.2 EMI/EMC Modeling: State of the Art

Current EMI/EMC modeling tools cannot do everything. That is, they cannot take the complete mechanical and electrical CAD files, compute overnight, and provide the engineer with a green/red light for pass/fail for the regulatory standard desired. The EMI/EMC engineer is needed to reduce the overall product into a set of problems that can be realistically modeled. The engineer must decide where the risks are in the product design and analyze those areas.

This means that the EMI/EMC engineer must remain an integral part of the EMI/EMC design process. Modeling will not replace the EMI/EMC engineer. Modeling is only one of the tools that EMI/EMC engineers have at their disposal. The knowledge and experience that the EMI/EMC engineer uses during the design process are needed to determine which area of the design needs further analysis and modeling.

Often the problem to be analyzed will require a multistage model. The results of one model's simulation will provide the input to the next stage model. This allows the model to be optimized for each particular portion of the problem, and the results combined. Thus much larger overall problems can be analyzed than by using a brute force approach, in which the entire problem is modeled at once. Again, the

EMI/EMC engineer needs to understand the problem and the modeling techniques well enough to know where to break it into individual simulations.

1.3 Tool Box Approach

No single modeling technique will be the most efficient and accurate for every possible model needed. Unfortunately, many commercial packages specialize in only one technique and try to force every problem into a particular solution technique. The EMI/EMC engineer has a wide variety of problems to solve, requiring an equally wide set of tools. The "right tool for the right job" approach applies to EMI/EMC engineering as much as it does to building a house or a radio. You would not use a putty knife to cut lumber, or a soldering iron to tighten screws, so why use an inappropriate modeling technique?

A wide range of automated EMI/EMC tools are available to the engineer. Automated tools include design rule checkers that check Printed Circuit Board (PCB) layout against a set of predetermined design rules; quasi-static simulators, which are useful for inductance/capacitance/resistance parameter extraction when the component is much smaller than a wavelength; quick calculators using closed-form equations calculated by computer for simple applications; full-wave numerical simulation techniques as described in this book; and expert-system tools, which provide design advice based on a predetermined set of conditions. It is clear that these different automated tools are applied to different EMI/EMC problems and at different times in the design process. This book will focus on the full-wave numerical modeling and simulation techniques and on how to apply these techniques to real-world EMI/EMC problems.

Different modeling techniques are suited to different problems. For example, the Method of Moments (MoM) technique is perfectly suited for a long wire simulation, since it only determines the currents on conductors, such as metal surfaces and wires, and it is independent of the volume of free space around the wires. However, the Finite Element Method (FEM) and the Finite-Difference Time-Domain (FDTD) method are not well suited to model long-wire simulations as in a computer with an external cable, since they require a volume of space to be modeled around the wire, and this volume must be large enough to have the computational domain boundary in the far field. Thus, using the FEM or the FDTD techniques for these applications results in a computationally inefficient model. On the other hand, there are problems for which the MoM is not a suitable choice, therefore, a set of tools that contain different modeling techniques is a great asset to the EMI/EMC engineer.

1.4 Brief Description of EMI/EMC Modeling Techniques

There are a variety of electromagnetic modeling techniques. Which is the "best" technique is cause for a significant amount of debate and often becomes a matter of which school the developer attended, and which technique his or her professor specialized in. Many of the techniques are specialized for certain configurations and require cumbersome tailoring when used for each problem. Some techniques are not particularly generic and require in-depth knowledge of electromagnetics and the modeling technique. Still others are useful only for far-field problems, such as determining a radar cross-section of a piece of military equipment. None of these specialized far-field techniques will be discussed here, since they have little use for the typical EMI/EMC engineer's problems.

Three techniques are typically used for EMI/EMC modeling problems: the FDTD technique, the MoM technique, and the FEM. Each technique will be briefly described here and then in greater detail in Chapters 3 through 5, so the EMI/EMC engineer can better understand how and when to use them.

1.4.1 Finite-Difference Time-Domain

The FDTD technique is a volume-based solution to Maxwell's differential equations. Maxwell's equations are converted to central difference equations and are solved directly in the time domain. The entire volume of space surrounding the object to be modeled must be gridded, usually into square or rectangular grids, small compared to the shortest wavelength of interest, and each grid location is identified as metal, air, or whatever material desired. Figure 1.1 shows an example of such a grid for a two-dimensional case. Once the grid parameters are established, the electric and magnetic fields are determined throughout the grid at a particular time. Time is advanced one time step, and the fields are

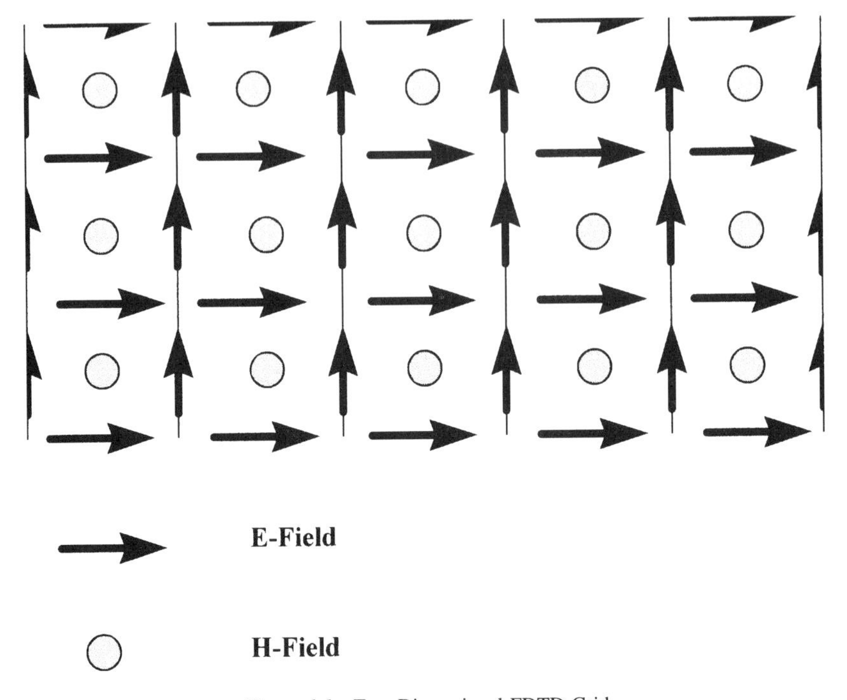

Figure 1.1 Two-Dimensional FDTD Grid

determined again. Thus, the electric and magnetic fields are determined at each time step based on the previous values of the electric and magnetic fields.

Once the fields have propagated thoughout the meshed domain, the FDTD simulation is complete and the broadband frequency response of the model is determined by performing a Fourier transform of the time domain results at the specified monitor points. Since the FDTD method provides a time domain solution, a wide band frequency domain result is available from a single simulation.

Since the FDTD technique is a volume-based solution,[3] the edges of the grid must be specially controlled to provide the proper radiation response. The edges are modeled with an Absorbing Boundary Condition (ABC). There are a number of different ABCs, mostly named after their inventors. In nearly all cases, the ABC must be electrically remote from the source and all radiation sources of the model, so that the far-

[3]The entire volume of the computational domain must be gridded.

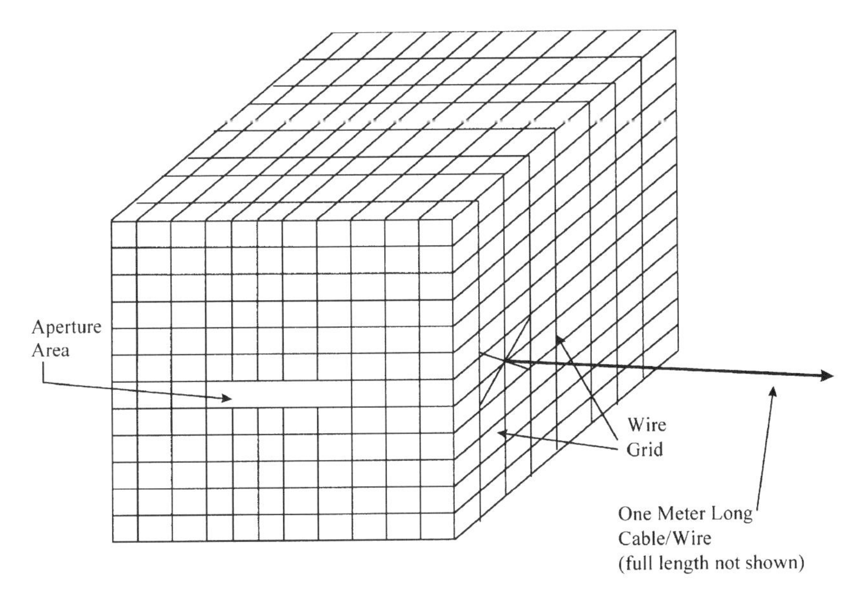

Figure 1.2 MoM Wire Mesh Model of Shielded Enclosure With 1-m-Long Cable Attached

field assumption of the ABC holds true and the ABC is reasonably accurate. Typically, a good ABC for the FDTD technique will provide an effective reflection of less than –60 dB.

Naturally, since the size of the gridded computational area is determined from the size of the model itself, some effort is needed to keep the model small. The solution time increases as the size of the computational area (number of grid points) increases. The FDTD technique is well suited to models containing enclosed volumes with metal, dielectric, and air. The FDTD technique is not well suited to modeling wires or other long, thin structures, as the computational area overhead increases very rapidly with this type of structure.

1.4.2 Method of Moments

The MoM is a surface current technique.[4] The structure to be modeled is converted into a series of metal plates and wires.[5] Figure 1.2 shows an example of a shielded box converted to a wire grid with a long

[4]Only the surface currents are determined, and the entire volume is not gridded.

[5]Often a solid structure is converted into a wire frame model, eliminating the metal plates completely.

attached wire. Once the structure is defined, the wires are broken into wire segments (short compared to a wavelength) and the plates are divided into patches (small compared to a wavelength). From this structure, a set of linear equations is created. The solution to this set of linear equations finds the RF currents on each wire segment and surface patch. Once the RF current is known for each segment and patch, the electric field at any point in space can be determined by solving for each segment/patch and performing the vector summation.

When using the MoM, the currents on all conductors are determined, and the remaining space is assumed to be air. This facilitates the efficiency of the MoM technique in solving problems with long thin structures, such as external wires and cables. Since the MoM technique finds the currents on the conductors, it models metals and air very efficiently. However, dielectric and other materials are difficult to model using MoM with standard computer codes.

The MoM technique is a frequency domain solution technique. Therefore, if the solution is needed at more than one frequency, the simulation must be run for each frequency. This is often required, since the source signals within the typical computer have fast rise times, and therefore wide harmonic content.

1.4.3 Finite Element Method

The FEM is another volume-based solution technique. The solution space is split into small elements, usually triangular or tetrahedral shaped, referred to as the finite element mesh. The field in each element is approximated by low-order polynomials with unknown coefficients. These approximation functions are substituted into a variational expression derived from Maxwell's equations, and the resulting system of equations is solved to determine the coefficients. Once these coefficients are calculated, the fields are known approximately within each element.

As in the above techniques, the smaller the elements, the more accurate the final solution. As the element size become small, the number of unknowns in the problem increases rapidly, thus increasing the solution time.

The FEM is a volume-based solution technique; therefore, it must have a boundary condition at the boundary of the computational space. Typically, the FEM boundaries must be electrically distant away from the structure being analyzed and must be spherical or cylindrical in shape. This restriction results in a heavy overhead burden for FEM users, since the number of unknowns is increased dramatically in comparison to other computational techniques.

1.5 Other Uses for Electromagnetics Modeling

Although this book will focus mostly on EMI/EMC modeling, and converting those types of problems into realistic models, there are many other uses for modeling. Antenna design, radar cross section, and microwave circuit analysis are only a few. These types of problems tend to have software focused especially on those problems; however, the techniques used for EMI/EMC modeling may be easily applied to these other specialized problems. In general, the most effective EMI/EMC modeling engineers will take an electromagnetic view of the overall problem, breaking it into source and receive, for analysis.

1.6 Summary

EMI/EMC problems are here to stay and are becoming more complex as personal communication devices proliferate and computer speeds continue to rise. Every electronic product on the market today and planned for tomorrow requires EMI/EMC considerations. Those engineers who insist on performing product design using previous methods only will quickly find themselves with design projects that are too expensive, either because of overdesign or because of repeated design cycles before regulatory compliance is reached. Although not every design project, nor every EMI/EMC design feature must be modeled, modeling/simulation can be a very useful tool to engineers. Experience has shown that once the initial hesitation to use something new is overcome, engineers find ways to use the modeling tools that they had never previously imagined.

Chapter 2

Electromagnetic Theory and Modeling

2.1 Introduction

It is well established that electromagnetic theory forms the backbone of electromagnetic interference and compatibility work. A good understanding of electromagnetic theory is highly desirable; however, it does not by itself lead to an understanding of the complexities of the electromagnetic interference phenomenon. This is because the interactions of electromagnetic fields with complex objects, as would be the case in real applications, cannot simply be predicted without abstraction of reality or, in other words, without creating a *model* of the physical system being analyzed. Creating an *electromagnetic* model that faithfully resembles the electromagnetic field behavior in the physical world is a challenging exercise that requires basic understanding of electromagnetic theory but would also require a good understanding of circuit theory, and of course, a sense of which physical factors are relevant to the model.

The electromagnetic theory is one of the most concise theories of the physical sciences. It is a body of knowledge that centers around a set of four partial differential equations known as Maxwell's equations. The essence of the theory is determining or characterizing the relationship between the electric and magnetic fields in the presence of different media with different properties. It could be stated that the electromagnetic theory has reached an advanced level of maturity. Despite this, it is essential to recognize that the maturity of the theory does not necessarily imply a parallel maturity in understanding of its application to the ever-increasing complex circuits and systems that are the hallmark of today's electronics technology.

Maxwell's equations relate the electric **E** and magnetic **H** fields to their excitation source in the presence of structures composed of different materials. The **E** and **H** fields are physically unobservable, that is there is no device in existence that can measure the field magnitude at any point in space in a direct fashion. However, even if such a device exists, another complication arises in that the presence of any "measuring device" will perturb the field that is being sought in the first place. The electrical quantities that are physically observable are the voltage v and current i and these can be measured with great accuracy. Therefore, when working with an electromagnetic system, the presence of the **E** and **H** fields can only be gauged indirectly. The v and i are analogous, in many respects, to the **E** and **H** fields. The v and i are used to describe the electrical properties of a circuit that is assumed to be sizeless, whereas the **E** and **H** fields describe the electrical properties of a circuit that occupies a finite physical domain. When the circuit is no longer sizeless, it is no longer referred to as a circuit, but as an electromagnetic device or system. However, regardless of the size of the circuit, the values of interest that can be measured are v and i.

This chapter presents Maxwell's equations in their integral and differential forms. The connections between **E, H,** v, and i are established. The interdependence of all these becomes meaningful when the circuit grows to occupy a finite, yet small enough domain such that the transfer of energy across this domain can be considered instantaneous. The field behavior in these circuits is referred to as a quasi-static field. As the circuit, or electromagnetic system, grows larger in size such that a finite amount of time is needed for the energy to travel across it, the fields exhibit a propagating characteristic. These fields will be referred to as propagating, or radiating fields. Radiating fields establish the first linkage between circuits that are separated by electrically large distances.

Finally, the subject of modeling is introduced. In general, modeling is referred to as the process of simulating physical phenomenon using computers. Here a fine distinction is made between *numerical modeling* and *electromagnetic modeling.* Numerical modeling is the procedure by which an analytical equation is converted into a discrete, or algebraic system of equations that can be solved on a computer. Numerical modeling is a subject that is vast and applies to many disciplines in the physical sciences. By contrast, electromagnetic modeling is the exclusive domain of electromagnetic scientists and engineers where the physical environment, including the primary sources of radiation, is abstracted and converted to mathematical entities, in order to be part of, or compatible with, the numerical model of the governing equations.

The material presented in this chapter on electromagnetic field theory is intended to be a refresher and neither comprehensive nor introductory. The objective is to link the elements of modeling, including field theory, numerical analysis, and electromagnetic modeling, in a way that facilitates an understanding that allows the EMI/EMC practitioner to know how the relevant equations and paradigms fit together.

2.2 Time-Varying Maxwell's Equations

Maxwell's equations are a set of four partial differential equations that govern the time-space relationship between the electric $\mathbf{E}(t,x,y,z)$ and magnetic $\mathbf{H}(t,x,y,z)$ fields. These equations are expressed as:

$$\nabla \times \mathbf{H} = \varepsilon \frac{\delta \mathbf{E}}{\delta t} + \mathbf{J} \tag{2.1}$$

$$\nabla \times \mathbf{E} = -\mu \frac{\delta \mathbf{H}}{\delta t} \tag{2.2}$$

$$\nabla \cdot \varepsilon \mathbf{E} = \rho \tag{2.3}$$

$$\nabla \cdot \mu \mathbf{H} = 0 \tag{2.4}$$

In (2.1), $\mathbf{J}$ is the conduction current density given by

$$\mathbf{J} = \sigma \mathbf{E} \tag{2.5}$$

Equations (2.1) to (2.4) describe the relationship between **E** and **H** in the presence of a material characterized by three parameters: the permittivity, ε, the permeability μ and the conductivity σ. The first of these equations, (2.1) states that a magnetic field changing in time produces an electric field that changes, or varies in space. Similarly, (2.2) states that an electric field changing in time produces a magnetic field that varies in space. Also deduced from (2.1) and (2.2) is that a time-varying electric field produces a time-varying magnetic field, and vice-versa. Equation (2.3) relates the flux, $\varepsilon\mathbf{E}$ to electric charge density, ρ, and (2.4) is an affirmation of the absence of magnetic charge.

It is difficult to deduce further information about the relationship between **E** and **H,** as given by Maxwell's equations than we have thus far. However, more insight can be gained by considering the integral forms of Maxwell's equations. The integral forms corresponding to (2.1) and (2.2) are referred to as Faraday's and Ampere's laws, given, respectively, by:

$$\oint_c \mathbf{E} \cdot d\mathbf{L} = \int_S \mu \frac{\delta \mathbf{H}}{\delta t} \cdot d\mathbf{S} \tag{2.6}$$

$$\oint_c \mathbf{H} \cdot d\mathbf{L} = I + \int_S \varepsilon \frac{\delta \mathbf{E}}{\delta t} \cdot d\mathbf{S} \tag{2.7}$$

where I is a current that corresponds to the conduction current density **J.**

The integral forms corresponding to (2.3) and (2.4) are referred to as Gauss's law and are given by:

$$\int_S \varepsilon \mathbf{E} \cdot d\mathbf{S} = \int_V \rho \, dv \tag{2.8}$$

$$\int_S \mu \mathbf{H} \cdot d\mathbf{S} = 0 \tag{2.9}$$

In the above integrals, $\oint_c$ refers to integration along a closed contour *c*.

The integral forms of Maxwell's equations are useful in establishing the relation between fields and circuits. However, before this connection is established, the assumption of quasi-static fields should hold.

2.2.1 Quasi-Static Fields

The term *quasi-static* field behavior refers to the assumption that the energy transfer across a circuit is instantaneous. That is a change in one part of the circuit affects all other parts of the circuit instantaneously. A circuit that satisfies the quasi-static assumption is said to be *electrically small.* Under the quasi-static assumption, the integral forms of Maxwell's equations are directly used to determine the voltage and current from impressed electromagnetic fields.

Faraday's law shows that a change in the magnetic field across a surface generates an electric field along the contour that fully surrounds, or bounds, the surface. Suppose we have a surface enclosed by a wire with a resistor connecting the two ends of the wire, as shown in Figure 2.1. A time-varying magnetic field across the surface **S** produces a current in the wire that generates a voltage drop across the resistor R. This can be shown simply by enforcing (2.6):

$$\int_1^2 \mathbf{E} \cdot d\mathbf{L} + \int_2^1 \mathbf{E} \cdot d\mathbf{L} = \mu \int_S \frac{\delta \mathbf{H}}{\delta t} \cdot d\mathbf{S} \tag{2.10}$$

When the integration is performed counterclockwise, the first integral is performed over the wire. Since **E** is zero on a wire (assuming infinite conductivity), this integral is zero. Now assume the resistor occupies a finite length, then the second integral on the left-hand side is simply the voltage drop across the resistor. Therefore, we have

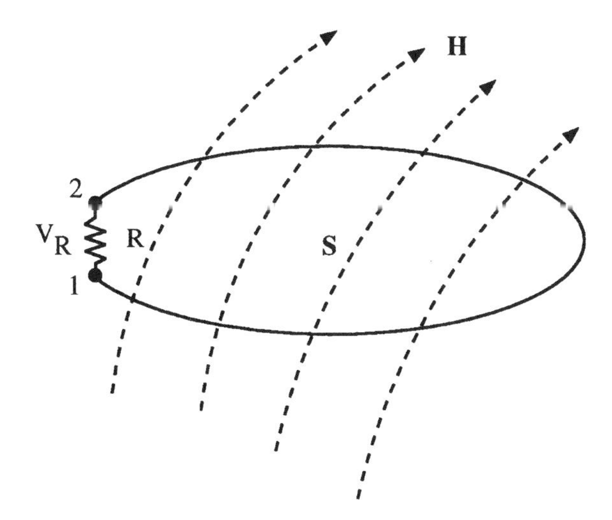

Figure 2.1 Demonstration of Faraday's Law

$$V_R = \int_S \mu \frac{\delta \mathbf{H}}{\delta t} \cdot d\mathbf{S} \tag{2.11}$$

Equation (2.11) establishes a link between the time-varying magnetic field and the voltage. (For completeness, one needs to account for the magnetic field that the current produces, however, this is considered negligible under the assumption of infinite conductivity.) Notice that if the field is not time-varying, the voltage drop is zero.

To develop the relationship between the fields and the current, Ampere's law is used. This law states that the line integration of the magnetic field along a contour is equal to the current that passes across the surface bounded by the contour. Consider Figure 2.2 which shows the same circuit shown in Figure 2.1, but with a closed contour that surrounds the wire. According to Ampere's law, as long as the contour **c** defining the surface **S** remains unchanged, the line integral of $\dot{\mathbf{H}}$ around the contour will be constant. Since the contour **c** shown in Figure 2.2 can form the boundary of an infinite number of surfaces, the only contribution from the right-hand side of (2.7) is *I,* which is the current that one measures in the wire. Thus, we have

$$\oint_C \mathbf{H} \cdot d\mathbf{L} = I = \frac{V_R}{R} \tag{2.12}$$

Now, if there is a capacitor forming a discontinuity in the wire, as shown in Figure 2.3, it is possible to have the surface bound by the

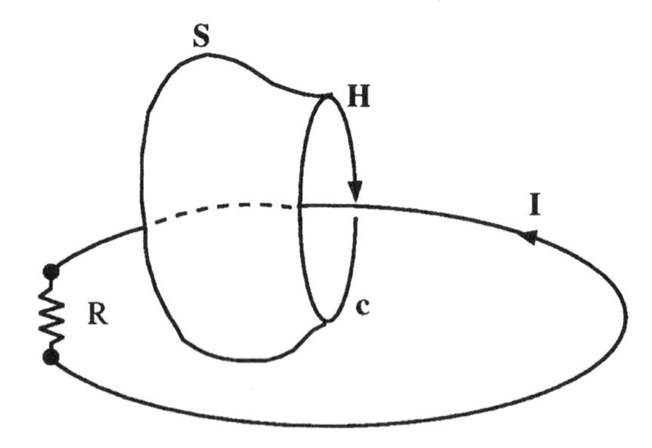

Figure 2.2 Demonstration of Ampere's Law, Showing Contribution of Conduction Current

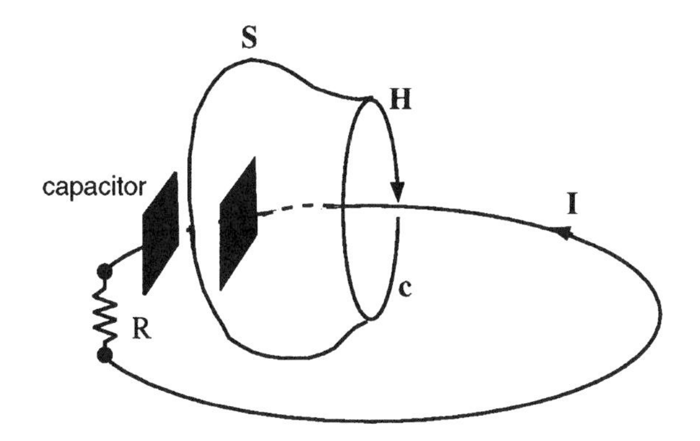

Figure 2.3 Demonstration of Ampere's Law, Showing Contribution of Displacement Current

contour to fall between the two plates of this capacitor. If the surface is chosen as in Figure 2.3, the only contribution to the left-hand side of (2.7) is the integral of $\varepsilon(\delta\mathbf{E}/\delta t)$, which gives the so-called displacement current.

When the circuit is no longer electrically small, a change in the magnetic field in part of the circuit (with respect to time) does affect other parts of the circuit; however, this effect takes place after a finite time. This implies that the field has to travel, or *propagate,* from one side to the other. Such a circuit need no longer be physically connected. The propagation of fields is the primary force behind the interference phenomenon. When propagation determines the characteristics of **E** and **H,** the circuit, or system, is referred to as an electromagnetic environment.

2.2.2 Radiating Fields

In free space, Maxwell's equations reduce to

$$\nabla \times \mathbf{H} = \varepsilon_0 \frac{\delta \mathbf{E}}{\delta t} \tag{2.13}$$

$$\nabla \times \mathbf{E} = -\mu_0 \frac{\delta \mathbf{H}}{\delta t} \tag{2.14}$$

$$\nabla \cdot \varepsilon_0 \mathbf{E} = 0 \tag{2.15}$$

$$\nabla \cdot \mu_0 \mathbf{H} = 0 \tag{2.16}$$

These source-free equations can be manipulated using vector identities to produce two equations that describe the behavior of either of **E** or **H** independently:

$$\nabla^2 \mathbf{E} = \mu_0 \varepsilon_0 \frac{\delta^2 \mathbf{E}}{\delta t^2} \tag{2.17}$$

$$\nabla^2 \mathbf{H} = \mu_0 \varepsilon_0 \frac{\delta^2 \mathbf{H}}{\delta t^2} \tag{2.18}$$

where ε_0 and μ_0 are the permittivity and permeability of free space.

These two equations are known as the *wave equations*. Solutions to wave equations are referred to as *waves*. That is, the **E** and **H** fields behave as waves that travel in space. Therefore, an excitation or an electromagnetic disturbance at point A causes a wave to travel away from the disturbance in a manner governed by the wave equations.

To illustrate the concept of waves, assume that the **E** field has only E_z as the only nonzero component. (The existence of only one field component can be supported by Maxwell's equations and is discussed later in the chapter.) Then (2.17) reduces to:

$$\frac{\delta^2 E_z}{\delta z^2} = \mu_0 \varepsilon_0 \frac{\delta^2 E_z}{\delta t^2} \tag{2.19}$$

or the more familiar form

$$\frac{\delta^2 E_z}{\delta z^2} - \frac{1}{c^2} \frac{\delta^2 E_z}{\delta t^2} = 0 \tag{2.20}$$

where $c = 1/(\mu_0 \varepsilon_0)^{-1/2}$ is the speed of light in free space.

The field solution is generally very complex and depends on the source of excitation and on the geometry of the problem. Regardless of how complex the field solution is, it must exhibit wave behavior. This dictates that the solution to E_z must be a combination of functions of the form:

$$E_z = f(z - ct) + g(z + ct) \tag{2.21}$$

The physical interpretation of (2.21) is that an electromagnetic disturbance that takes place at a point A in space propagates, or travels, to a point B with the speed of light. Suppose that points A and B are part of an electromagnetic environment such as a current source **J** and a wire loop separated by the distance R as illustrated in Figure 2.4. If this environment is electrically small, by definition, the effect of turning the source **J** on would create a voltage drop V_R instantaneously, which means that no propagation takes place. Now if the distance R is large enough (depending on the dimensions and temporal excitation of the source), the effect of the source takes finite time to reach the wire loop. When it takes finite time for energy to reach the loop, the environment is said to exhibit propagating, or radiating fields.

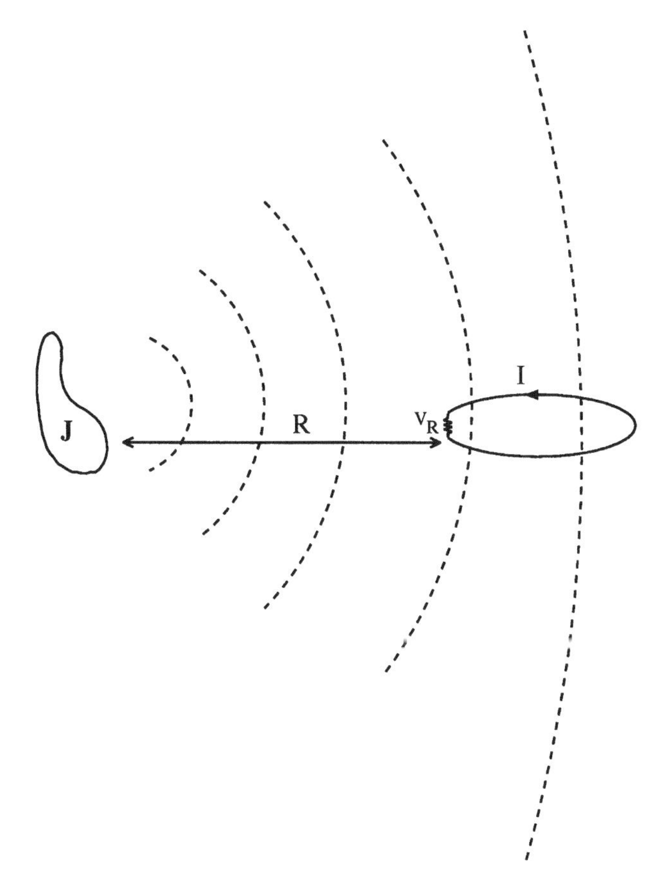

Figure 2.4 A Current Distribution **J** Affecting a Remote Circuit Causing a Current and a Voltage Drop in the Circuit

2.3 Field Solutions Using Potentials

Maxwell's equations are coupled first-order partial differential equations that can be solved directly to determine the fields due to an excitation current source at any point in space. The direct analytical solution of Maxwell's equations, however, can only be found for a small class of problems that have limited practical application. To extend the application of the equations to a wider class of problems, use is made of scalar and vector potentials, which are mathematical functions that have a convenient dependence on the excitation current.

Suppose there exists a charge density ρ and current density $\mathbf{J}_e$, the scalar ϕ and vector potentials $\mathbf{A}$ satisfy the following two partial differential equations:

$$\nabla^2 \mathbf{A} - \mu\varepsilon \frac{\delta^2 \mathbf{A}}{\delta t^2} = -\mu \mathbf{J}_e \tag{2.22}$$

$$\nabla^2 \phi - \mu\varepsilon \frac{\delta^2 \phi}{\delta t^2} = -\frac{\rho}{\varepsilon} \tag{2.23}$$

The relationship between **E, H, A** and ϕ is governed by the following equations:

$$\mathbf{E} = -\nabla\phi - \frac{\delta \mathbf{A}}{\delta t} \tag{2.24}$$

$$\mathbf{H} = \frac{1}{\mu}\nabla \times \mathbf{A} \tag{2.25}$$

$$\nabla \cdot \mathbf{A} = -\mu\varepsilon \frac{\delta\phi}{\delta t} \tag{2.26}$$

Equations (2.22) and (2.23) are enforced at a single point in space having the current excitation $\mathbf{J}_e$. If a current $\mathbf{J}$ occupies a volume V in space as shown in Figure 2.5 it can be shown (using Green's theorem) that the potential can be expressed as a continuous summation or integral of the charge and current distributions:

$$\phi = \frac{\mu}{4\pi}\int_V \frac{\rho_{ret}}{R}\, dv' \tag{2.27}$$

$$\mathbf{A} = \frac{\mu}{4\pi}\int_V \frac{\mathbf{J}_{ret}}{R}\, dv' \tag{2.28}$$

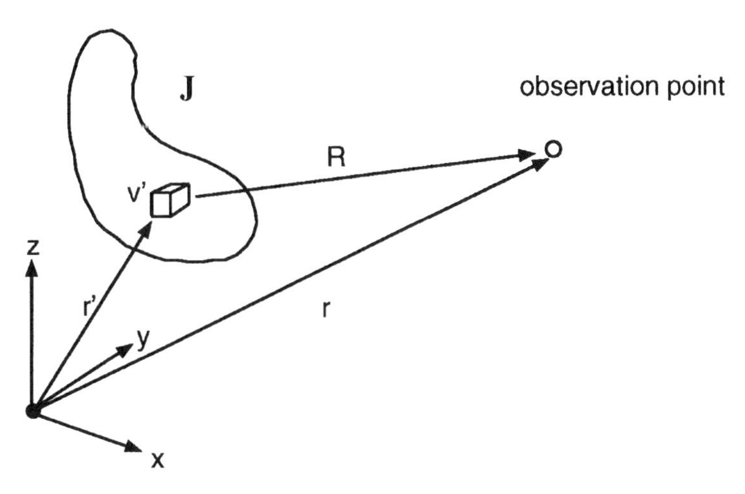

Figure 2.5 Volume Current Distribution in Space

where R is the distance from the current distribution to the observation point. The subscript *ret* denotes the fact that the charge and current are evaluated at the retarded time $t - R/c$, where c is the speed of light in the medium, given by $1/(\mu\varepsilon)^{-1/2}$.

2.4 Maxwell's Equations in Frequency Domain

The time-varying Maxwell's equations fully describe the behavior of the **E** and **H** fields from the moment the source excitation is turned on. The temporal waveform of the source excitation can take on any shape, and it can still be described by Maxwell's equations. In many electromagnetic applications the interest lies in the steady-state response of the system or environment to a sinusoidal temporal excitation of the form $\sin(\omega t)$, where ω is the frequency of oscillation. Under such excitation, and after the fields reach their steady-state behavior, all the fields behave in an oscillatory manner and are referred to as *time – harmonic* or *frequency – domain* fields. These fields can be expressed as:

$$\mathbf{E}(t,x,y,z) = \tilde{\mathbf{E}}(x,y,z)\ cos(\omega t) \tag{2.29}$$

$$\mathbf{H}(t,x,y,z) = \tilde{\mathbf{H}}(x,y,z)\ cos(\omega t) \tag{2.30}$$

Using phasor notation, we have:

$$\mathbf{E}(t,x,y,z) = \tilde{\mathbf{E}}(x,y,z)\ cos(\omega t) = Re\{\tilde{\mathbf{E}}(x,y,z)e^{j\omega t}\} \tag{2.31}$$
$$\mathbf{H}(t,x,y,z) = \tilde{\mathbf{H}}(x,y,z)\ cos(\omega t) = Re\{\tilde{\mathbf{H}}(x,y,z)e^{j\omega t}\} \tag{2.32}$$

where the tilde (˜) indicates a phasor form of the field, that is, the part of the field that contains the spatial variation only.

Maxwell's curl equations simplify to:

$$\nabla \times \tilde{\mathbf{H}} = j\omega\varepsilon\tilde{\mathbf{E}} + \tilde{\mathbf{J}} \tag{2.33}$$
$$\nabla \times \tilde{\mathbf{E}} = -j\omega\mu\tilde{\mathbf{H}} \tag{2.34}$$

The time-harmonic fields are very useful in facilitating simple analyses of many problems. This is particularly the case when the source of excitation of the electromagnetic environment, such as a radiating antenna, has a single frequency of oscillation. However, since any temporal waveform can be represented using Fourier theorem as a summation of sinusoids, frequency-domain analysis can be equally valuable in studying time-varying fields.

Extending the time-harmonic analysis to the scalar ϕ and vector $\mathbf{A}$ potentials, equations (2.22), (2.23), and (2.26) transform to the following:

$$\nabla^2\tilde{\mathbf{A}} + \mu\varepsilon\omega^2\tilde{\mathbf{A}} = -\mu\tilde{\mathbf{J}}_e \tag{2.35}$$
$$\nabla^2\tilde{\phi} + \mu\varepsilon\omega^2\tilde{\phi} = -\frac{\rho}{\varepsilon} \tag{2.36}$$
$$\nabla \cdot \tilde{\mathbf{A}} = -j\omega\mu\varepsilon\tilde{\phi} \tag{2.37}$$

The $\tilde{\mathbf{E}}$ field can be found directly from (2.35) to (2.37) as:

$$\tilde{\mathbf{E}} = -\frac{j}{\omega\mu\varepsilon}\nabla(\nabla \cdot \tilde{\mathbf{A}}) - j\omega\tilde{\mathbf{A}} \tag{2.38}$$

While (2.38) appears to complicate the formulation instead of simplifying it, the usefulness of this formulation becomes clear by considering the application of the vector potential $\tilde{\mathbf{A}}$ to a current source oriented, or polarized in the z-direction, J_z. Then, $\tilde{\mathbf{A}}$ will correspondingly have only a z-component given by:

$$A_z = \frac{\mu}{4\pi} \int_{V'} J_z \frac{e^{-jkR}}{R} dv' \qquad (2.39)$$

It follows that the $\tilde{\mathbf{E}}$ field components can directly be related, through an integral equation, to the current. For instance, the E_z component is given by:

$$E_z = \frac{-j}{4\pi} \int_{V'} k^2 \frac{e^{-jkR}}{R} + \frac{\partial^2}{\partial z^2} \frac{e^{-jkR}}{R} dv' \qquad (2.40)$$

It will be shown in Chapter 4 that this expression for the field leads to the development of the Method of Moments.

2.5 Electromagnetic Fields in Two-Dimensional Space

Maxwell's equations in three dimensions describe the field behavior in three-dimensional physical space. Any physical phenomenon including radiating electromagnetic fields takes place in physical space. Two-dimensional space, described by two independent space variables x and y, is a mathematical, or nonphysical space. However, Maxwell's equations in two-dimensional space have unique features because they reduce to special forms that allow simpler and faster solution and analysis than would be required in three-dimensional space.

Before commenting on the meaning of Maxwell's equations in two-dimensional space, the characteristics of these equations in this non-physical space must be explored. We consider a two-dimensional space in x and y. This implies that the field exhibits no variation in the z-direction (zero partial derivative with respect to z). Expanding Maxwell's curl equations in free space, (2.1) gives:

$$\frac{\partial H_z}{\partial y} = \varepsilon \frac{\partial E_x}{\partial t} \qquad (2.41)$$

$$-\frac{\partial H_z}{\partial x} = \varepsilon \frac{\partial E_y}{\partial t} \qquad (2.42)$$

$$\left(\frac{\partial H_y}{\partial x} - \frac{\partial H_x}{\partial y}\right) = \varepsilon \frac{\partial E_z}{\partial t} \qquad (2.43)$$

And, (2.2) gives:

$$\frac{\partial E_z}{\partial y} = -\mu \frac{\partial H_x}{\partial t} \tag{2.44}$$

$$\frac{\partial E_z}{\partial x} = \mu \frac{\partial H_y}{\partial t} \tag{2.45}$$

$$\left(\frac{\partial E_y}{\partial x} - \frac{\partial E_x}{\partial y} \right) = -\mu \frac{\partial H_z}{\partial t} \tag{2.46}$$

Because of this unique partial decoupling of Maxwell's equations in two dimensions, it is possible to support the existence of only three field components at any time. Suppose E_x and E_y are zero, Maxwell's equations reduce to:

$$\left(\frac{\partial H_y}{\partial x} - \frac{\partial H_x}{\partial y} \right) = \varepsilon \frac{\partial E_z}{\partial t} \tag{2.47}$$

$$\frac{\partial E_z}{\partial y} = -\mu \frac{\partial H_x}{\partial t} \tag{2.48}$$

$$\frac{\partial E_z}{\partial x} = \mu \frac{\partial H_y}{\partial t} \tag{2.49}$$

When only E_z, H_x, and H_y exist, satisfying (2.47) to (2.49), the fields are referred to as Transverse Magnetic (TM) fields, or TM polarization fields, implying that the magnetic fields lie only in the transverse, or (x,y) plane.

In a fully analogous fashion, when H_x and H_y are zero, then Maxwell's equations reduce to:

$$\frac{\partial H_z}{\partial y} = \varepsilon \frac{\partial E_x}{\partial t} \tag{2.50}$$

$$-\frac{\partial H_z}{\partial x} = \varepsilon \frac{\partial E_y}{\partial t} \tag{2.51}$$

$$\left(\frac{\partial E_y}{\partial x} - \frac{\partial E_x}{\partial y} \right) = -\mu \frac{\partial H_z}{\partial t} \tag{2.52}$$

Similarly, when only H_z, E_x, and E_y exist satisfying (2.50) to (2.52), the fields are referred to as Transverse Electric (TE) fields, or TE polarization fields, implying that the electric fields lie only in the transverse, or $x - y$ plane.

When the fields are either TM or TE, the determination of a single field component (E_z in the TM case and H_z in the TE case) is sufficient to fully describe the remaining field components, as can be seen from the above equations. This implies that the field in either of these two cases can be specified by a single partial differential equation. In fact, this can also be inferred directly from (2.17) and (2.18). For instance, in the case of the TM polarization, (2.17) reduces to

$$\nabla^2 E_z - \mu_0 \varepsilon_0 \frac{\partial^2 E_z}{\partial t^2} = 0 \qquad (2.53)$$

In Frequency domain analysis, (2.53) reduces to:

$$\nabla^2 \tilde{E}_z + k^2 \tilde{E}_z = 0 \qquad (2.54)$$

where $k = \omega\sqrt{\mu_0 \varepsilon_0}$ is the wave number. Equation (2.54) is known as the Helmholz wave equation.

When the rate of change of the field with respect to time is small, the second term of (2.53) is assumed to be negligible and (2.53), in the frequency domain, reduces to:

$$\nabla^2 \tilde{E}_z = 0 \qquad (2.55)$$

Equation (2.55) is known as the Laplace equation, which is useful to approximate quasi-static field behavior.

In certain electromagnetic applications, such as in the theory and application of waveguides, it is possible to have a current source that generates fields that support either TE or TM behavior. This is typically dictated by the form of excitation used. However, in many other applications, the reduction of the three-dimensional Maxwell's equations to the reduced two-dimensional form is a matter of convenience and practicality. For instance, take a plane wave that is perpendicular to the x-y plane and incident on a cylindrical structure with a uniform cross section parallel to the z-axis as shown in Figure 2.6. If this cylindrical structure is very long (electrically), and assuming that the two ends of the structure do not affect the fields significantly, the

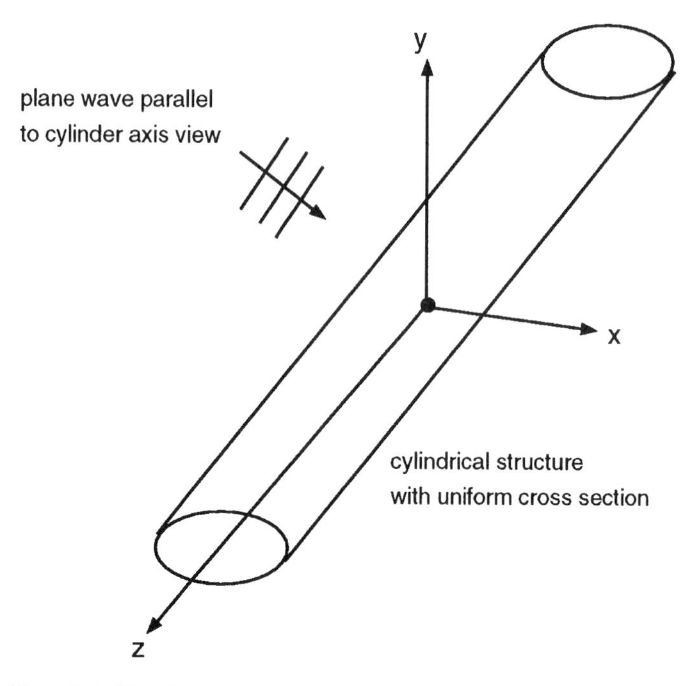

Figure 2.6 Plane Wave Incident on an Electrically Long Cylindrical Structure

interaction of the wave with the structure can be studied by considering the reduced two-dimensional problem shown in Figure 2.7.

The reduction of the order of the problem, such as assuming certain physical conditions that allow for the reduction of Maxwell's equations from three-dimensions to two-dimensions, can be very helpful in many

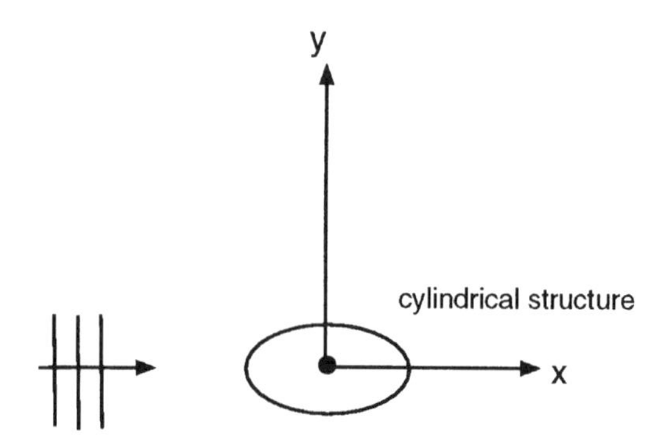

Figure 2.7 Two-Dimensional Abstraction of the Cylinder Problem

problems. It can give the modeler a feeling of how certain fields can affect a structure without resorting to the more complex three-dimensions physical model. However, three words of caution. First, reduction of Maxwell's equations to two dimensions is valid under certain conditions and, most of the time, under certain assumptions relating to the field behavior. Second, the source of energy can be a crucial factor in determining whether a three- to two-dimension reduction is valid. Third, the final results of the analysis must be interpreted very carefully. In some problems, correlations of the two-dimensional field behavior to the true physical field behavior are possible after a transformation in variables. In other instances, there exists no physical correspondence to the solution. However, this does not lessen the important insight that can be gained, even if no direct physical field correlation can be made.

2.6 Numerical Modeling

When modeling an electromagnetics system, the solution of Maxwell's equations is, in most cases, the ultimate objective. Maxwell's equations are in essence a coupled system of partial differential equations having two unknown functions: the electric **E** and magnetic **H** fields. The unknown functions are defined over a domain (the domain or space of the problem) subject to boundary conditions. The initial source of **E** and **H** is a primary current source that is considered the cause of radiation. This current source is independent of the resulting fields and is also independent of the structure and material of the medium.

If an analytical solution for the system of coupled equations exists, subject to specified boundary conditions, our task is lessened considerably. Unfortunately, Maxwell's equations are sufficiently complex that their analytical solution exists only for very simple cases. For instance, analytical solutions can be directly obtained for an infinitesimal current element radiating in free space, or an infinitely long line source parallel to an infinitely circular cylinder. The problems that can be solved analytically are referred to as canonical problems, or problems having separable geometries. Unfortunately, these problems are of very limited scope and are hardly existent in a real-world application. This limitation creates a fundamental necessity to solve Maxwell's equations using numerical techniques.

The field of applied mathematics is replete with numerical techniques that can be used to solve partial differential equations. These numerical techniques vary in complexity, required computer resources, and finally, they can vary considerably in their solution time. These methods include the Finite-Difference Frequency-Domain (FDFD), Finite-Difference Time-Domain (FDTD), Finite Element Method (FEM), Method of Moments (MoM), and Boundary Element Method. Some of these techniques, such as the FDFD and FDTD, solve Maxwell's equations directly, meaning that the analytical partial differential equations of Maxwell are transformed into discretized equations. Other techniques, such as the FEM and the MoM, first transform Maxwell's equations into a different form, (integral form) that is solved indirectly in which an intermediate solution is obtained before the unknown **E** and **H** are determined.

Irrespective of the technique used, a numerical procedure entails the discretization of the fields (the unknown function) over the space of the problem. By the discretization of the field, it is meant that the fields in space are defined only at a numerable or *discrete* set of points, or grid. This step is the nexus of numerical techniques. Once this is done, the continuous field description is replaced by a discrete distribution. Now instead of seeking an analytical solution or a continuous field across the domain, we seek a function whose unknown variables are the field values at the discrete points in space.

There is always a furious debate amongst computational scientists and engineers as to which method is the best to use. In most cases, the bias toward a certain technique depends on the level of familiarity and comfort one has with that technique. To rank numerical techniques in terms of strength can be a superfluous exercise. This is because the types of problems that arise in electromagnetics, and especially in EMI/EMC work, are very diverse. A method that might be very efficient to solve a particular problem, can be very memory intensive when addressing a totally different problem. Second, an important consideration before selecting a method is the determination in advance of the objective of the model. This can be very helpful in determining, for instance, the suitability of either time-domain or frequency-domain techniques.

The selection of a numerical technique to address a specific problem requires in most cases careful adherence to the limitations imposed by the parameters of the technique. The most significant parameters relate to the density of the discretization in space which specifies the separation between two field points in space. These parameters affect the accuracy of the solution in what is referred to as discretization errors. In certain numerical techniques, the discretization errors affect not only the accuracy of the solution, but also the field characteristics, such as field dispersion.

Numerical techniques that solve the time-varying fields have unique features that distinguish them from frequency domain techniques. For instance, stability of the numerical method refers to whether the solution is sensitive to numerical artifacts, such as computer precision, or to other numerical artifacts that can be part of the numerical techniques itself or of the electromagnetic model.

2.7 Electromagnetic Modeling

Numerical modeling can be a subject of concern to computational scientists. By contrast, electromagnetic modeling requires diverse engineering skills typically gained through experience. Electromagnetic modeling may be defined as the transformation of the physical device into a representative geometry. The geometry is then described by space and material parameters which should be compatible with the numerical model chosen for the analysis and should reflect the physical characteristics of the device as accurately as possible. Second, the energy source that causes the initial excitation or disturbance in the system needs to be parameterized, or modeled into a mathematical entity called a numerical source. This numerical source must be compatible with the numerical model and, more importantly, should reflect, as faithfully as possible, the actual physical source. Third, the complex electromagnetic environment in which the device operates need to be converted and simplified into a mathematical model.

These two steps typically involve simplifications and assumptions that are problem dependent. For instance, in many practical problems, a good conducting structure having a very high, yet finite, conductivity can be assumed to have infinite conductivity to allow for the enforcement of simple boundary conditions. In certain critical applications, such as in the study of cavity resonance, the assumption of infinite conductivity might prove invalid.

In certain applications, a plate with a small thickness can be assumed to have zero thickness, whereas in other applications, the finite thickness of the plate can affect field behavior significantly, especially near the edges.

The accurate parameterization and modeling of materials can be crucial in the study of energy absorption in the human head due to the presence of a cellular phone. For example, studies have shown that a simplified model of human flesh, blood and tissue can lead to inaccurate results showing significant variation in energy absorption from experimental data.

Source modeling plays a unique role in electromagnetic applications and modeling. The types of electromagnetic problems, including those encountered in EMI/EMC applications, can be classified into two categories. The first category includes the types of problems in which the interest lies in characterizing the behavior of a device or environment. For instance, in studying the effectiveness of shields to electromagnetic waves, one is interested in the relative performance of a shield with a certain design and geometry in comparison to a shield having a different design. In the shielding study, the way in which the electromagnetic fields were initially generated is of minimal importance as long as these waves are present with certain characteristics (plane waves, TM waves, or TE waves). The second category of problems includes problems where the interest lies in determining the absolute radiated power that emanates from a certain device, such as a VLSI package. In the first category, source modeling can be performed using a variety of ways, however, in the second category, the success of the simulation highly depends on the implementation of a source model that faithfully resembles the physical source of excitation. As will be explained in Chapters 6 and 7, what makes a good source model depends to a large extent on the engineer's experience, knowledge, and previous modeling experience.

In general, electromagnetic modeling is not an exact science. A good model depends on the engineers' understanding of the physics of the problem, and also the various simplifications that are permissible in a certain and unique application. Finally, it should be stressed that effective modeling is a process that is inseparable from the numerical code or solver used to obtain the final analysis.

2.8 Summary

The objective of this chapter is to give a brief summary of the equations that describe field behavior. This summary was not intended to be a comprehensive exposition of the preliminaries of electromagnetic theory, but to highlight the equations that are of pertinent interest to the field of electromagnetic modeling. More specifically, the presentation provides the basis for the next three chapters where the computational techniques are developed.

References

1. J.D. Jackson, *Classical Electrodynamics,* John Wiley & Sons, New York, 1962.
2. E.C. Jordan and K.G. Balmain, *Electromagnetic Waves and Radiating Systems,* Prentice-Hall, Englewood Cliffs, N.J., 1968.
3. W.L. Stutzman and G.A. Thiele, *Antenna Theory and Design,* John Wiley & Sons, New York, 1981.
4. S. Ramo and J.R. Whinnery, *Fields and Waves in Modern Radio,* John Wiley & Sons, New York, 1944.
5. W.H. Hayt, Jr., *Engineering Electromagnetic,* McGraw-Hill, New York, 1981.
6. R.F. Harrington, *Time-Harmonic Electromagnetic Fields,* McGraw-Hill, New York, 1961.
7. C.A. Balanis, *Advanced Engineering Electromagnetics,* John Wiley & Sons, New York, 1989.
8. L.C. Shen and J.A. Kong, *Applied Electromagnetics,* PWS Engineering, Boston, 1987.
9. R.E. Collin, *Field Theory of Guided Waves,* McGraw-Hill, New York, 1960.
10. R.C. Booton, Jr., *Computational Method for Electromagnetics and Microwaves*, John Wiley & Sons, New York, 1992.

Chapter 3

The Finite-Difference Time-Domain Method

3.1 Introduction

The Finite-Difference Time-Domain (FDTD) method provides a direct integration of Maxwell's time-dependent equations. During the past decade, the FDTD method has gained prominence amongst numerical techniques used in electromagnetic analysis. Its primary appeal is its remarkable simplicity. Furthermore, since the FDTD is a volume-based method, it is exceptionally effective in modeling complex structures and media. However, the distinct feature of the FDTD method, in comparison to the Method of Moments (MoM) and the Finite Elements Method (FEM) (see Chapters 4 and 5) is that it is a time-domain technique. This implies that one single simulation results in a solution that gives the response of the system to a wide range of frequencies. The time-domain solution, represented as a temporal waveform, can then be decomposed into its spectral components using Fourier Transform techniques. This advantage makes the FDTD especially well-suited for most EMI/EMC problems in which a wide frequency range is intrinsic to the simulation.

The FDTD method is a volume-based method requiring dividing the space of the solution into a uniform mesh composed of cells. Over each cell, the **E** and **H** field components will be defined. This aspect of the FDTD method is identical to the FEM. However, in the FEM, a matrix equation is developed, which can then be solved in a variety of ways. In the FDTD method, no matrix solution is needed. Instead, the **E** and **H** fields are staggered in space, and the leapfrog in time method is employed. This allows a direct solution of the fields, in *time*. In other words, as time evolves, the solution for each field component is determined for that particular instant in time and then stored in memory.

The development of the FDTD here will be based on the Yee cell. The special feature of the Yee cell is that the **E** and **H** field components are staggered one half space-cell apart, which facilitates differencing schemes that are sufficiently accurate, as will be discussed below.

The FDTD method is perhaps the simplest and least complex of numerical techniques, however, a word of caution to the practical modeler: While the FDTD method is simple and its pertinent equations are easy to encode, an effective and useful simulation using FDTD depends on numerical simulation aspects that are not necessarily intrinsic to the FDTD formulation. Most specifically, a successful FDTD simulation depends on (1) accurate numerical modeling of the primary source of energy, (2) accurate mesh truncation techniques to prevent the presence of spurious waves in the solution domain, and (3) accurate and reliable field extension formulation to allow for calculation of the field in the region exterior to the simulation domain of the FDTD method. The FDTD basic formulation, in addition to these important factors that affect the simulation, are the subjects of this chapter. Finally, an overview is given to highlight the relative significance of the errors that arise from each of the steps comprising the entire FDTD simulation.

3.2 Two-Dimensional FDTD

The reduction of Maxwell's equations to two dimensional space is useful for obtaining the solutions of many problems. Most likely, however, the two-dimensional equations are used where the assumption is made that the field is invariant in one of the three spatial dimensions. Once the development of the FDTD in two-dimensional space is understood, the generalization to the three-dimensional space will be easier to follow.

Assuming field invariance in the z-direction, Maxwell's equations reduce to two possible sets of equations. Throughout the discussion, the invariance is taken with respect to the z-direction, and hence reference will be made to either Transverse Electric (TE) or Transverse Magnetic (TM) polarization (see Chapter 2) without reference to the direction of invariance.

For TM polarization, Maxwell's equations reduce to

$$-\mu\frac{\partial H_x}{\partial t} = \frac{\partial E_z}{\partial y} \tag{3.1}$$

$$\mu\frac{\partial H_y}{\partial t} = \frac{\partial E_z}{\partial x} \tag{3.2}$$

$$\frac{\partial E_z}{\partial t} = \frac{1}{\varepsilon}\left(\frac{\partial H_y}{\partial x} - \frac{\partial H_x}{\partial y} - \sigma E_z\right) \tag{3.3}$$

For the TE polarization, we have

$$\frac{\partial E_x}{\partial t} = \frac{1}{\varepsilon}\left(\frac{\partial H_z}{\partial y} - \sigma E_x\right) \tag{3.4}$$

$$\frac{\partial E_y}{\partial t} = -\frac{1}{\varepsilon}\left(\frac{\partial H_z}{\partial x} + \sigma E_y\right) \tag{3.5}$$

$$\frac{\partial H_z}{\partial t} = \frac{1}{\mu}\left(\frac{\partial E_x}{\partial y} - \frac{\partial E_y}{\partial x}\right) \tag{3.6}$$

It is important to keep in mind, that these equations are intended to approximate two unique and totally different physical situations that directly depends on the polarization of the radiated field.

A direct solution to the systems in (3.1) to (3.3) and in (3.4) to (3.6) entails a transformation of the differential equations to a set of difference equations using a central difference approximation scheme to the spatial and temporal derivatives.

The entire space of the problem is then filled with a uniform mesh. The **E** and **H** field components are positioned in a staggered configuration as shown in Figure 3.1 for the TM polarization case (the roles of the **E** and **H** field components are reversed for the TE case). Since the mesh is uniform in the *x*- and *y*-directions, the location of the field is identified by the indeces *i* and *j*. This amounts to evaluating the fields at a set of discrete points in space. Similarly, the time scale is descretized

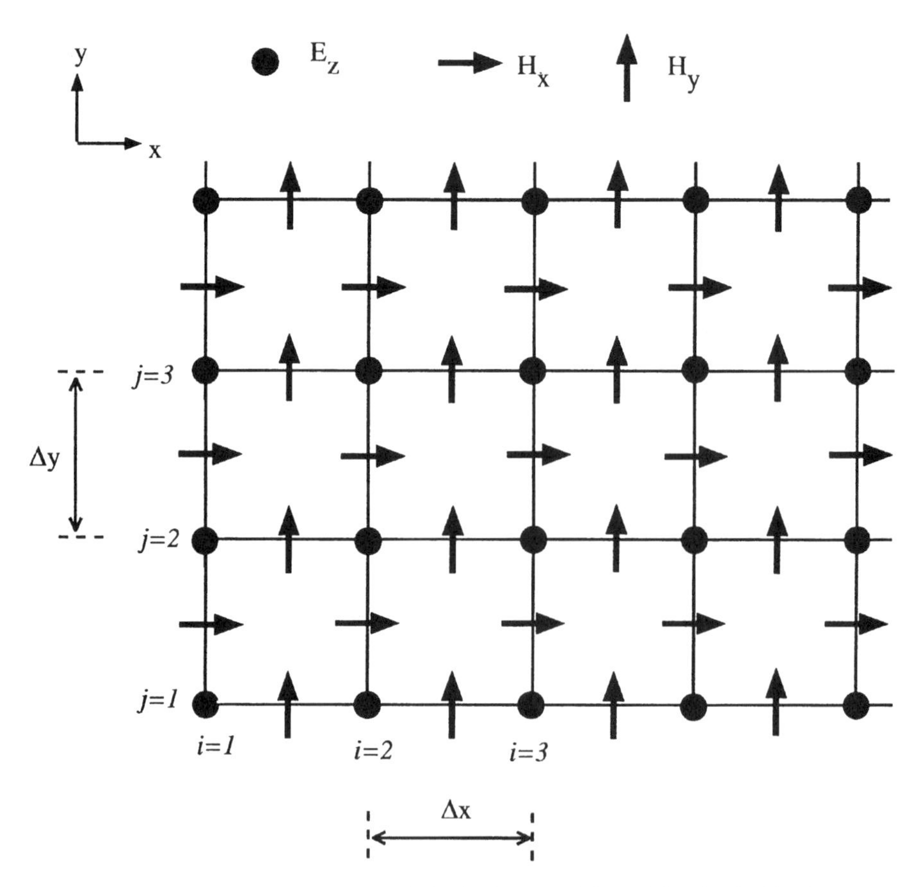

Figure 3.1 Two-Dimensional FDTD Mesh for the Transverse Magnetic Polarization Case Showing the Staggering of the **E** and **H** Fields One Half Space-Cell Apart

into uniform time steps, spaced by a time interval referred to as the time step, Δt.

We shall denote the field at the spatial location ($i\Delta x$, $j\Delta y$) and time step $n\Delta t$ by

$$\mathbf{E}(i\Delta x, j\Delta y, n\Delta t) = \mathbf{E}^n(i, j) \tag{3.7}$$

The first fundamental step in the FDTD method is using central-differencing approximation for the spatial derivatives. This gives a differencing scheme that is second-order accurate. For the system in (3.1) to (3.3), we obtain:

$$\left.\frac{\partial E_z}{\partial y}\right|_{y=j\Delta y} \rightarrow \frac{E_z(i, j+1) - E_z(i, j)}{\Delta y} \tag{3.8}$$

$$\left.\frac{\partial E_z}{\partial x}\right|_{x=i\Delta x} \rightarrow \frac{E_z(i+1, j) - E_z(i, j)}{\Delta x} \tag{3.9}$$

Similar difference equations are applied to $\delta H_y/\delta x$ and $\delta H_x/\delta y$.

The next fundamental step is to use the central difference approximation, as above, to approximate the temporal derivative. What gives the FDTD method its most distinguishing feature, however, is that the **E** and **H** fields are evaluated not at the same time instant, but at two points in time separated by one half-time step. For the system in (3.1) to (3.3), we obtain

$$\left.\frac{\partial H_x}{\partial t}\right|_{t=n\Delta t} \rightarrow \frac{H_x^{n+\frac{1}{2}} - H_x^{n-\frac{1}{2}}}{\Delta t} \tag{3.10}$$

$$\left.\frac{\partial H_y}{\partial t}\right|_{t=n\Delta t} \rightarrow \frac{H_y^{n+\frac{1}{2}} - H_y^{n-\frac{1}{2}}}{\Delta t} \tag{3.11}$$

and for $\delta E_z/\delta t$, we obtain

$$\left.\frac{\partial E_z}{\partial t}\right|_{t=n\Delta t} \rightarrow \frac{E_z^{n+1} - E_z^n}{\Delta t} \tag{3.12}$$

The staggering of the field components one half-cell apart in space and at time instances that are one half-step apart in space allows for the complete and sequential evaluation of the fields as time advances

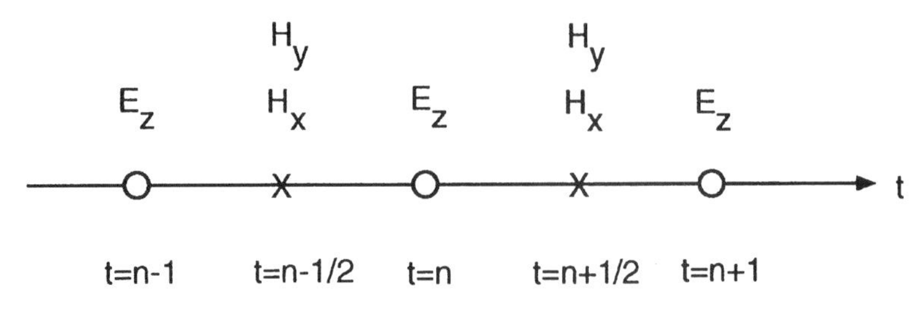

Figure 3.2 Leapfrog Scheme Used in FDTD

from the start of the simulation to its completion. This method of staggering the **E** and **H** field components in time is known as the leapfrog scheme and is summarized in Figure 3.2.

Finally, applying the above approximations to the TM polarization equations, we have

$$H_x^{n+\frac{1}{2}}(i, j) = H_x^{n-\frac{1}{2}}(i, j) - \frac{\Delta t}{\mu_{ij}\Delta y}\left[E_z^n(i, j+1) - E_z^n(i, j)\right] \quad (3.13)$$

$$H_y^{n+\frac{1}{2}}(i, j) = H_y^{n-\frac{1}{2}}(i, j) + \frac{\Delta t}{\mu_{ij}\Delta x}\left[E_z^n(i+1, j) - E_z^n(i, j)\right] \quad (3.14)$$

$$E_z^{n+1}(i, j) = E_z^n(i, j) + \frac{\Delta t}{\varepsilon_{ij}\Delta x}\left[H_y^{n+\frac{1}{2}}(i+1, j) - H_y^{n+\frac{1}{2}}(i, j)\right] - \frac{\Delta t}{\varepsilon_{ij}\Delta y}\left[H_x^{n+\frac{1}{2}}(i, j+1) - H_x^{n+\frac{1}{2}}(i, j)\right] \quad (3.15)$$

Similarly, for the TE polarization case, we have

$$E_x^{n+\frac{1}{2}}(i, j) = E_x^{n-\frac{1}{2}}(i, j) + \frac{\Delta t}{\varepsilon_{ij}\Delta y}\left[H_z^n(i, j+1) - H_z^n(i, j)\right] \quad (3.16)$$

$$E_y^{n+\frac{1}{2}}(i, j) = E_y^{n-\frac{1}{2}}(i, j) - \frac{\Delta t}{\varepsilon_{ij}\Delta x}\left[H_z^n(i+1, j) - H_z^n(i, j)\right] \quad (3.17)$$

$$H_z^{n+1}(i, j) = H_z^n(i, j) + \frac{\Delta t}{\mu_{ij}\Delta y}\left[E_x^{n+\frac{1}{2}}(i, j+1) - E_x^{n+\frac{1}{2}}(i, j)\right] - \frac{\Delta t}{\mu_{ij}\Delta x}\left[E_y^{n+\frac{1}{2}}(i+1, j) - E_y^{n+\frac{1}{2}}(i, j)\right] \quad (3.18)$$

where ε_{ij} and μ_{ij} correspond to the permittivity and permeability of each cell in the mesh.

3.3 Three-Dimensional FDTD

The discussion of the two-dimensional FDTD scheme highlights the essential fundamentals of the FDTD technique, especially the staggering of the **E** and **H** grid and the use of central differencing to approximate the differential operators. In three-dimensional space, the extension is analogous; instead of a two-dimensional mesh, we develop a three-dimensional mesh composed of the Yee cell as its basic building block. The field will then be identified by the three indexes i, j, k in a manner similar to (3.7). The Yee cell in three-dimensional space is shown in Figure 3.3, where the field components are staggered in space as before, such that they are a half-cell apart. Using the fundamental approximations for the FDTD in (3.8) to (3.11), Maxwell's equations transform to the following discrete finite-difference equations:

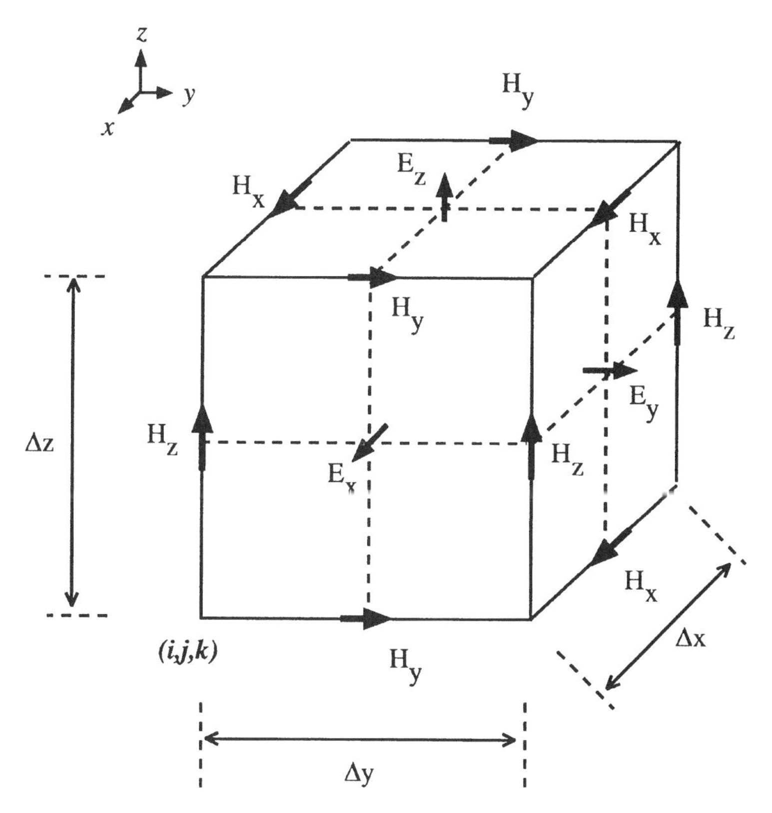

Figure 3.3 Yee Cell in Three-Dimensional Space

$$H_x^{n+\frac{1}{2}}(i,j,k) = H_x^{n-\frac{1}{2}}(i,j,k) + \frac{\Delta t}{\mu_{ijk}}\left[\frac{E_y^n(i,j,k+1) - E_y^n(i,j,k)}{\Delta z} - \frac{E_z^n(i,j+1,k) - E_z^n(i,j,k)}{\Delta y}\right] \tag{3.19}$$

$$H_y^{n+\frac{1}{2}}(i,j,k) = H_y^{n-\frac{1}{2}}(i,j,k) + \frac{\Delta t}{\mu_{ijk}}\left[\frac{E_z^n(i+1,j,k) - E_z^n(i,j,k)}{\Delta x} - \frac{E_x^n(i,j,k+1) - E_x^n(i,j,k)}{\Delta z}\right] \tag{3.20}$$

$$H_z^{n+\frac{1}{2}}(i,j,k) = H_z^{n-\frac{1}{2}}(i,j,k) + \frac{\Delta t}{\mu_{ijk}}\left[\frac{E_x^n(i,j+1,k) - E_x^n(i,j,k)}{\Delta y} - \frac{E_y^n(i+1,j,k) - E_y^n(i,j,k)}{\Delta x}\right] \tag{3.21}$$

$$E_x^{n+1}(i,j,k) = E_x^n(i,j,k) + \frac{\Delta t}{\varepsilon_{ijk}}\left[\frac{H_z^{n+\frac{1}{2}}(i,j+1,k) - H_z^{n+\frac{1}{2}}(i,j,k)}{\Delta y} - \frac{H_y^{n+\frac{1}{2}}(i,j,k+1) - H_y^{n+\frac{1}{2}}(i,j,k)}{\Delta z}\right] \tag{3.22}$$

$$E_y^{n+1}(i,j,k) = E_y^n(i,j,k) + \frac{\Delta t}{\varepsilon_{ijk}}\left[\frac{H_x^{n+\frac{1}{2}}(i,j,k+1) - H_x^{n+\frac{1}{2}}(i,j,k)}{\Delta z} - \frac{H_z^{n+\frac{1}{2}}(i+1,j,k) - H_z^{n+\frac{1}{2}}(i,j,k)}{\Delta x}\right] \tag{3.23}$$

$$E_z^{n+1}(i,j,k) = E_z^n(i,j,k) + \frac{\Delta t}{\varepsilon_{ijk}}\left[\frac{H_y^{n+\frac{1}{2}}(i+1,j,k) - H_y^{n+\frac{1}{2}}(i,j,k)}{\Delta x} - \frac{H_x^{n+\frac{1}{2}}(i,j+1,k) - H_x^{n+\frac{1}{2}}(i,j,k)}{\Delta y}\right] \tag{3.24}$$

The order for calculating the fields is as before, as illustrated in Figure 3.4. First, the electric field components are calculated, then time is advanced one half-step, then the magnetic field is updated from the previously calculated electric field and so on.

Because the FDTD method is a time-domain technique, the time iteration must conform to causality principles that imply that the progression, or advance, from one node to the next, in any direction, does not exceed the speed of light. It can be shown that this physical constraint

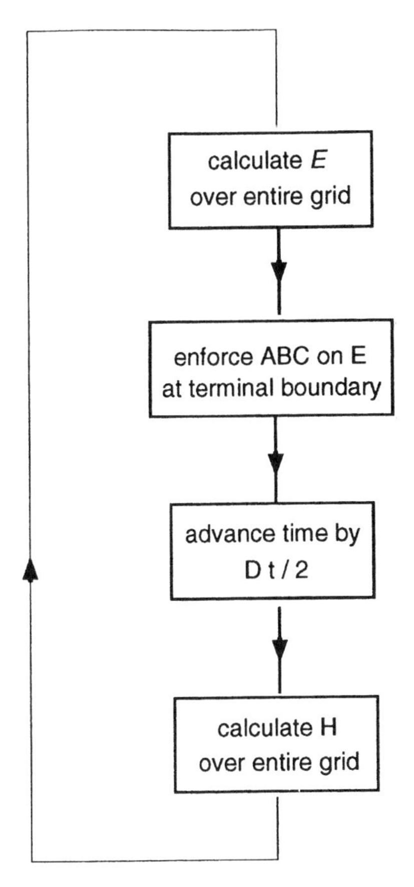

Figure 3.4 Sequence of Field Calculation Within the FDTD Algorithm

has a limiting effect on the relationship between the spatial steps Δx, Δy, Δz and the time step Δt. This constraint, known as the Courant's condition depends on the FDTD cell size and is given for two-dimensional and three-dimensional spaces respectively as

$$\Delta t < \frac{1}{v\sqrt{\frac{1}{\Delta x^2} + \frac{1}{\Delta y^2}}} \tag{3.25}$$

$$\Delta t < \frac{1}{v\sqrt{\frac{1}{\Delta x^2} + \frac{1}{\Delta y^2} + \frac{1}{\Delta z^2}}} \tag{3.26}$$

where v is the speed of light in the medium. The constraints in (3.25) and (3.26) apply to the medium with highest density, as that is where the speed of light is slowest.

3.4 Modeling Primary Radiation Sources

The FDTD method gives a highly accurate solution to Maxwell's equations. In classical electromagnetic scattering problems where the scattering response of an object needs to be determined, the exact nature of the source of energy plays a minor role in the simulation. EMI/EMC problems, however, differ markedly in that the source of energy, or the coupling of radio frequency (RF) energy into the structure/system plays an important role in determining the specific quantity of radiation that the device emits.

In numerical techniques, energy can be coupled into the system in a variety of ways. What is important to note here is that each numerical source may or may not correspond to a physical excitation. Therefore, the final response of the system depends, in addition to the physical structure and the medium, on the source coupling mechanism, illustrated in the diagram shown in Figure 3.5.

The most common source used to simulate a localized primary source is the impressed current, **J**, which is added to the displacement current in Maxwell's equations, giving:

$$\nabla \times \mathbf{H} = \varepsilon_0 \frac{\partial \mathbf{E}}{\partial t} + \mathbf{J} \tag{3.27}$$

For clarity, we assumed that the conductivity of the medium is zero, hence the absence of the conduction current density term in (3.27). Equation (3.27) is transformed to an FDTD difference equation in a manner consistent with the fundamental approximations discussed earlier. For instance, assuming a z-polarized current source in Transverse

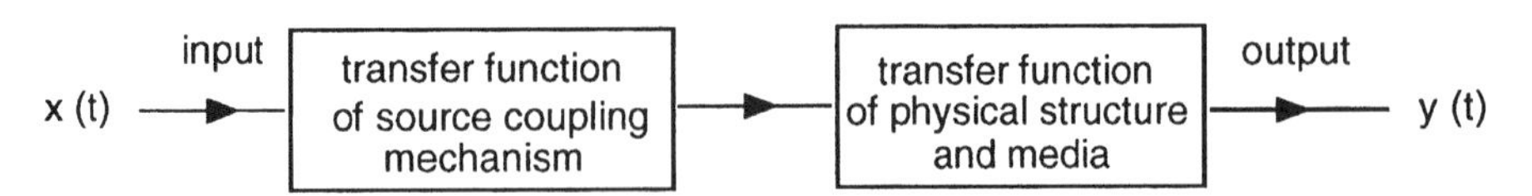

Figure 3.5 Source Coupling Transfer Functions

Magnetic (TM) polarization problem, the update equation for E_z becomes

$$E_z^{n+1}(i, j) = E_z^n(i, j) + \frac{\Delta t}{\varepsilon_{ijk}}\left[\frac{H_y^{n+\frac{1}{2}}(i+1, j) - H_y^{n+\frac{1}{2}}(i, j)}{\Delta x} - \frac{H_x^{n+\frac{1}{2}}(i, j+1) - H_x^{n+\frac{1}{2}}(i, j)}{\Delta y}\right] - \frac{\Delta t}{\varepsilon_{ij}} J_z^n(i, j) \tag{3.28}$$

Note that the current source is located at the same spatial point as the electric field; however, it coincides with the same time point as the magnetic field. The value given to J_z is a time waveform that depends on the problem at hand (see Chapter 6 for more details).

The impressed current source corresponds to different physical realizations depending on the polarization of the source. For instance, in the TE polarization case, the impressed current corresponds to an x- or y-directed current source, which can be interpreted as a physical current sheet that is normal to the $x - y$ plane. For TM polarization, the impressed current represents a point source in the computational plane. However, the physical correspondence would be an infinite line source that is normal to the computational plane. These are important considerations for the modeler to consider when using the different sources available in two-dimensional space.

In three-dimensional space, there is less ambiguity to what is meant by an impressed current source. For instance, a one cell long impressed current source corresponds to a Hertzian dipole.

In some applications, the impressed current is intended to simulate the response of a circuit. If the impressed current source was used by itself, the circuit's effect on the excitation are absent from the simulation. For instance, if the source of energy has a finite input impedance, after the input excitation pulse is switched off, the current source resembles an open circuit that may not be the desired effect. To remedy this problem, a finite impedance source can be used in FDTD simulations. This is accomplished by satisfying Ohm's law at the FDTD cell where the impedance source is desired.

To incorporate the resistive voltage source in an FDTD scheme, we consider a unit Yee cell in three-dimensional space. Considering a source polarized in the z-direction, and assuming a desired resistance of R, the cell is filled with a conductive material that has a value that corresponds to the resistance:

$$\sigma_{ijk} = \frac{\Delta z}{R \Delta x \Delta y} \tag{3.29}$$

Next, an impressed current is injected into the conductive cell. This current must satisfy Ohm's law in the cell. Maxwell's equation for the the $\mathbf{E}_z$ field becomes

$$\varepsilon_{ijk} \frac{\partial \mathrm{E}_z}{\partial t} = \left(\frac{\partial \mathrm{H}_y}{\partial x} - \frac{\partial \mathrm{H}_x}{\partial y} \right) + \frac{V_z}{R} \tag{3.30}$$

where V_z indicates a voltage source oriented in the z-direction. The addition of the current term to the field update equation in FDTD is performed by first using a time-average of $\mathbf{E}_z$ at $n + 1$ and n, in order to give a consistent and stable solution. Thus, we have

$$\begin{aligned} &\mathrm{E}_z^{n+1}(i, j, k) = \\ &A \mathrm{E}_z^{n}(i, j, k) + B \left[\frac{\mathrm{H}_y^{n+\frac{1}{2}}(i+1, j, k) - \mathrm{H}_y^{n+\frac{1}{2}}(i, j, k)}{\Delta x} \right. \\ &\left. - \frac{\mathrm{H}_x^{n+\frac{1}{2}}(i, j+1, k) - \mathrm{H}_x^{n+\frac{1}{2}}(i, j, k)}{\Delta y} \right] \\ &+ C V_z^{n+\frac{1}{2}} \end{aligned} \tag{3.31}$$

where

$$A = \frac{1 - \dfrac{\Delta t \Delta z}{2 R_z \varepsilon_0 \Delta x \Delta y}}{1 + \dfrac{\Delta t \Delta z}{2 R_z \varepsilon_0 \Delta x \Delta y}}$$

$$B = \frac{\dfrac{\Delta t}{\varepsilon_0}}{1 + \dfrac{\Delta t \Delta z}{2 R_z \varepsilon_0 \Delta x \Delta y}}$$

$$C = \frac{\dfrac{\Delta t}{R_z \varepsilon_0 \Delta x \Delta y}}{1 + \dfrac{\Delta t \Delta z}{2 R_z \varepsilon_0 \Delta x \Delta y}}$$

3.5 Numerical Dispersion and Anisotropy

Once Maxwell's equations are transformed into difference equations through the FDTD method, the dispersion relationship is altered in a manner that reflects the non-infinitesmal discretization of the FDTD scheme. The dispersion relationship for the fields derived from the analytical expression of Maxwell's equations in two-dimensional space is given by:

$$\frac{\omega^2}{c^2} = k_x^2 + k_y^2 \tag{3.32}$$

However, once the field is expressed over a discrete grid, the analyticity of the dispersion relationship is consequently altered. In the two-dimensional TM case, it can be shown that the dispersion relationship corresponding to the non-infinitesimal discretization with equal space steps in the x- and y-directions, is given by

$$\left(\frac{\Delta x}{c\Delta t}\right)^2 \sin^2(\omega\Delta t/2) = \sin^2\frac{\Delta x k_x^d}{2} + \sin^2\frac{\Delta x k_y^d}{2} \tag{3.33}$$

where k_x^d and k_y^d are the new wavenumbers corresponding to the discretized domain. Notice that this deviation in the dispersion relationship is purely due to grid effects, referred to as numerical artifact, or numerical dispersion. In models where the field is allowed to propagate for a distance comparable or greater than the longest wavelength, numerical dispersion can cause up to 10% error in the solution. However, in most EMI/EMC problems, the extensive propagation of the field is very uncommon, and even if the field needs to be evaluated at a distance from the radiating device, field extension techniques are used (to be discussed below), which pre-empt the emergence of dispersion errors.

Also, as a consequence of using a structured grid composed of the Yee cell, a constant phase front cannot travel in all directions in a uniform manner. This variation in the solution is referred to as grid anisotropy. This form of error can be appreciable only if a significant amount of energy travels in the diagonal direction of the grid.

As can be seen from the limit as the FDTD cell shrinks in size, grid dispersion and anisotropy diminish. Except for modeling problems requiring exceptionally high level of accuracy, grid anisotropy and dispersion are of lesser concern than other errors, such as those introduced by mesh truncation techniques.

3.6 Mesh Truncation Techniques

The FDTD is a volume-based method that requires discretization of the entire domain over which the solution is to be sought. For problems where radiation takes place in an open region, as the case in the largest percentage of EMI/EMC problems, the domain of the solution fills the entire space and is infinite in extent. To discretize such space is impossible because of the finite memory capability of computers. Therefore, the mesh has to be truncated to a finite size that can be manipulated with reasonable computer storage. Once the infinite space of the open region problem is truncated to a finite size, a mesh truncation formulation must be enforced on the outer boundaries of the computational domain in order to simulate the nonreflective nature of open space. The mesh terminal boundaries must be as nonreflective as possible, mimicking the behavior of free space, or in analogy to EMI/EMC testing, to a good anechoic chamber.

The computational domain size is directly proportional to the memory requirements, and because of this, it is advantageous to keep the boundaries which terminate the computational domain as close to the radiating structure as possible. With such constraint comes the challenge of designing a mesh truncation technique that is effective in absorbing:

1. Waves traveling at a multitude of angles
2. Evanescent waves
3. Fields with frequencies spanning a wide range

A poor mesh truncation technique can render the FDTD simulation useless, emphasizing that the quality of the FDTD analysis is highly dependent on the quality of the mesh truncation technique employed.

Mesh truncation schemes fall into three different categories: (1) exact or nonlocal boundary condition; (2) local absorbing boundary operators, or absorbing boundary conditions; and (3) absorbing material. Absorbing Boundary Conditions (ABCs) are mathematical constructions that take the form of differential operators which approximate the behavior of outgoing waves. Absorbing material is numerical media (physical or nonphysical) containing lossy material that works to gradually annihilate the fields as it penetrates through them.

The performance of mesh truncation techniques vary considerably, but what is important to keep in mind is that the accuracy that can be achieved is typically directly proportional to the overhead resources (memory and operation count) required by the technique. A large number of mesh truncation techniques are based on either mathematical or physical principles. In this section, we present three different effective techniques that vary in their performance in accordance with the overhead needed to implement the method. The exact nonlocal boundary condition will not be presented since its excessive memory and computational overhead makes it highly impractical.

3.6.1 Higdon's Absorbing Boundary Conditions

When an open region is terminated into a finite size, the terminal boundary can intersect either the **E** or **H** fields. Therefore, a boundary condition needs to be applied on only the field that lies along the terminal boundary. Typically, the boundary condition is applied at the **E** fields. In three-dimensional space, the **E** field, which is tangential to the boundary has two components and the boundary operator is applied to both of them. Let us consider the left-hand side of the terminal boundary shown in Figure 3.6. Higdon's first order boundary condition is given by

$$\left(\partial_x + \frac{\cos\phi}{c}\partial_t + \alpha\right)\mathbf{E} = 0 \tag{3.34}$$

where **E** is the field component which is tangential to the boundary, c is the speed of light, and α is a constant that depends on the grid size that is essential for maintaining a stable solution.

Higdon's boundary operators are the simplest boundary operators capable of completely absorbing waves traveling across the boundary at an angle ϕ with respect to the normal. It is a simple exercise to show that for a wave traveling perpendicular to the terminal boundary, having a velocity c, will be completely absorbed. However, for a wave traveling at an angle, $\theta \neq \phi$ from the x-axis, the terminal boundary generates a reflection that is dependent on the angle ϕ. Assuming sinusoidal excitation, that is an incident field having $e^{j\omega t}$ dependence, the reflection

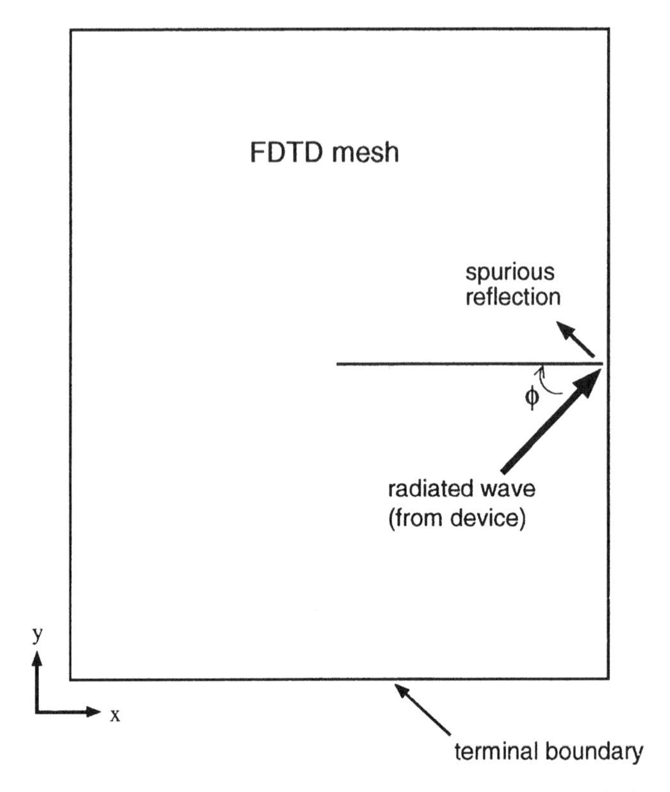

Figure 3.6 Artificial Terminal Boundary to Limit the Computational Domain in an FDTD Simulation

coefficient can be found by postulating incoming and outgoing waves in the computational domain:

$$E = e^{-jk_x x - jk_y y - jk_z z + j\omega t} + Re^{jk_x x - jk_y y - jk_z z + j\omega t} \tag{3.35}$$

then enforcing (3.34) on E, we have

$$R = (-1)\left(\frac{-jk_x + jk\cos\phi + \alpha}{jk_x + jk\cos\phi + \alpha}\right) \tag{3.36}$$

Here we note that since a typical radiating field contains a multitude of waves traveling at different angles, the reflection coefficient, R, varies from its lowest value to its highest as the angle of incidence increases from normal to grazing incidence.

A higher accuracy operator can be developed by cascading two first-order operators:

$$\left(\partial_x + \frac{\cos\phi_1}{c}\partial_t + \alpha_1\right)\left(\partial_x + \frac{\cos\phi_2}{c}\partial_t + \alpha_2\right)\mathrm{E} = 0 \tag{3.37}$$

giving yet a smaller reflection coefficient:

$$R = \left(\frac{-jk_x + jk\cos\phi + \alpha_1}{jk_x + jk\cos\phi + \alpha_1}\right)\left(\frac{-jk_x + jk\cos\phi + \alpha_2}{jk_x + jk\cos\phi + \alpha_2}\right) \tag{3.38}$$

Consequently, an Nth order operator can be constructed in the same manner, we have:

$$\prod_{i=1}^{N}\left(\partial_x + \frac{\cos\phi_i}{c}\partial_t + \alpha_i\right)\mathrm{E} = 0 \tag{3.39}$$

Here N designates the order of the operator. Clearly, the higher the order, the lower the reflection coefficient R. The corresponding R is given by

$$R = (-1)\prod_{i=1}^{N}\frac{-jk_x + jk\cos\phi_i + \alpha_i}{jk_x + jk\cos\phi_i + \alpha_i} \tag{3.40}$$

To incorporate (3.39) into the FDTD scheme, the differential operators have to be transformed into difference operators. This is performed by the following special transformations, which involve an averaging with respect to the secondary variable to guarantee the stability of the simulation:

$$\frac{\partial \mathrm{E}}{\partial t} \rightarrow \left[\frac{\mathrm{E}^{n+\frac{1}{2}}(i, j, k) - \mathrm{E}^{n-\frac{1}{2}}(i, j, k)}{\Delta t}\right]$$

$$\left[\frac{\mathrm{E}^{n+\frac{1}{2}}(i, j, k) + \mathrm{E}^{n+\frac{1}{2}}(i-1, j, k)}{2}\right] \tag{3.41}$$

Similarly,

$$\frac{\partial E}{\partial x} \rightarrow \left[\frac{E^{n+\frac{1}{2}}(i, j, k) - E^{n+\frac{1}{2}}(i-1, j, k)}{\Delta x}\right]$$

$$\left[\frac{E^{n+\frac{1}{2}}(i, j, k) + E^{n-\frac{1}{2}}(i, j, k)}{2}\right] \qquad (3.42)$$

With this transformation, (3.39) reduces to:

$$\prod_{i=1}^{N}\left[E_{i,j,k}^{n+\frac{1}{2}} + a_i E_{i-1,j,k}^{n+\frac{1}{2}} + b_i E_{i,j,k}^{n-\frac{1}{2}} + c_i E_{i-1,j,k}^{n-\frac{1}{2}}\right] = 0 \qquad (3.43)$$

where

$$a_i = \frac{-1 + \dfrac{\Delta x}{c\Delta_t} + \dfrac{\alpha_i \Delta_x}{2}}{1 + \dfrac{\Delta_x}{c\Delta_t} + \dfrac{\alpha_i \Delta_x}{2}}$$

$$b_i = \frac{1 - \dfrac{\Delta x}{c\Delta_t} + \dfrac{\alpha_i \Delta_x}{2}}{1 + \dfrac{\Delta_x}{c\Delta_t} + \dfrac{\alpha_i \Delta_x}{2}}$$

$$c_i = \frac{-1 - \dfrac{\Delta x}{c\Delta_t} + \dfrac{\alpha_i \Delta_x}{2}}{1 + \dfrac{\Delta_x}{c\Delta_t} + \dfrac{\alpha_i \Delta_x}{2}}.$$

As can be seen from (3.40), it is possible to optimize Higdon's Nth order ABC to favor the absorption of waves having a predetermined angle of incidence. This, however, takes away from the generality and robustness of the ABC, and it is recommended that ϕ_i be set to 0.

A special feature of Higdon's operator is its uniaxial nature, which means that the discretization of the operator involves field components that lie along a normal to the terminal boundary. This feature allows simple and effective treatment of corner regions that are typically responsible for a sizable spurious reflection.

Higdon's ABCs are an excellent representation of most boundary operators, and their flexibility, i.e., the choice of the order of the operator, allows the modeler to control the accuracy by the most suitable choice of the operator's order. It should be noted, however, that while the theoretical reflection coefficient of the operator decreases as the order increases, this cannot be maintained indefinitely since the higher the order, the higher the memory requirements and operation count. Furthermore, higher-order operators exceeding 4th order become harder to stabilize. The first and second order operators are stable, without the addition of the constant α_i, however, for 3rd and 4th order operators, an α_i in the range $0.005 \le \alpha_i\Delta_t/2 \le 0.01$ is sufficient to stabilize the solution in two-dimensional problems, while $0.01 \le \alpha_i\Delta_x/2 \le 0.05$ is adequate for three-dimensional problems.

3.6.2 Complementary Operators Method

The basic premise of the Complementary Operators Method (COM) is the cancellation of the first-order reflection that arises from the truncation of the computational domain. This cancellation is made possible by averaging two independent solutions of the problem. These two solutions are obtained by imposing boundary operators that are complementary to each other, in the sense that the errors generated by the two operators are equal in magnitude but 180 degrees out of phase. As a result of the averaging process, the first order reflections, consisting of either evanescent or traveling waves, are annihilated.

From the theory of Higdon's operators, we can derive two complementary operators by applying the two operators δx and δt independently on Higdon's Nth order to give the following two operators:

$$B_N^+ \mathrm{E} = \partial_x \prod_{i=1}^{N-1} (\partial_x + \frac{\cos\phi_i}{c}\partial_t + \alpha_i)\mathrm{E} = 0 \tag{3.44}$$

and

$$B_N^- \mathrm{E} = \partial_x \prod_{i=1}^{N-1} (\partial_x + \frac{\cos\phi_i}{c}\partial_t + \alpha_i)\mathrm{E} = 0 \tag{3.45}$$

We denote the corresponding reflection coefficients respectively, by R^+ and R^-, which are given, respectively, by

$$R^{+} = R[B_N^{-}] = -\prod_{i=1}^{N}\left(\frac{-jk_x + jk\cos\phi_i + \alpha_i}{jk_x + jk\cos\phi_i + \alpha_i}\right) \tag{3.46}$$

$$R^{-} = R[B_N^{+}] = \prod_{i=1}^{N}\left(\frac{-jk_x + jk\cos\phi_i + \alpha_i}{jk_x + jk\cos\phi_i + \alpha_i}\right) \tag{3.47}$$

Notice that $R^{+} = -R^{-}$. Thus averaging the two solutions obtained from applying each of the two operators separately, gives a solution containing only second-order reflections.

The implementation of the complementary operators method in an FDTD code takes two different forms, depending on the dimension of the space. In two-dimensional space, corner reflections constitute the second most dominant reflections because they reach the observation point faster than multiple reflections due to the scatterer. To cancel these reflections, four solutions instead of two need to be averaged, with each requiring an independent simulation. For each simulation, one needs to impose a unique combination of B_N^{-} and B_N^{+} over the four sides of the outer boundary.

The implementation of the COM requires splitting the FDTD computational space into two regions: a boundary layer and an interior, as shown in Figure 3.7. The interior region includes the scattering object and any localized sources. In the boundary layer, instead of defining one storage location for each of E_z, H_x, and H_y (as in the TM polarization case, for example), we allocate four storage locations to each field. Within the interior region, only a single value for each of the field components is stored as in typical FDTD implementation.

The next step is to average the four values obtained for each field at an interface that is placed immediately to the inside of the boundary layer. This averaging is performed at each time step. The width of the boundary layer must be wider than the stencil needed to discretize the boundary operators in (3.44) or (3.45).

The implementation of each of the two operators (3.46) and (3.47) in an FDTD scheme parallels the implementation of Higdon's ABC.

The extension of the COM to three-dimensional space is identical to the two-dimensional case, except that the annihilation of corner reflections would need a total of eight storage locations for each field component in the boundary layer. This can add a substantial memory

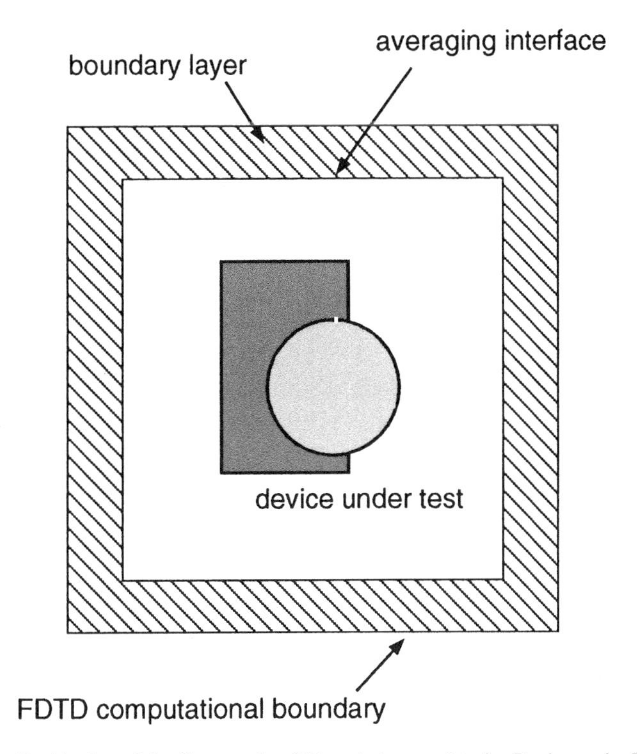

Figure 3.7 Partitioning of the Computational Domain into an Interior Region and a Boundary Layer for the Implementation of the Complementary Operators Method

overhead for moderate size problems. Therefore, the annihilation of reflections from side boundary can be performed with reasonable memory and computational overhead.

3.6.3 Perfectly Matched Layer

The concept of using a matched medium as a mesh truncation technique perhaps precedes the development of absorbing boundary operators. The idea was originally conceived based on the physical concept of impedance matching in circuits to minimize reflections. To illustrate this, let us consider first an interface between two media: a free-space and a medium with electric and magnetic conductivities, as in Figure 3.8. For TE polarization, we have:

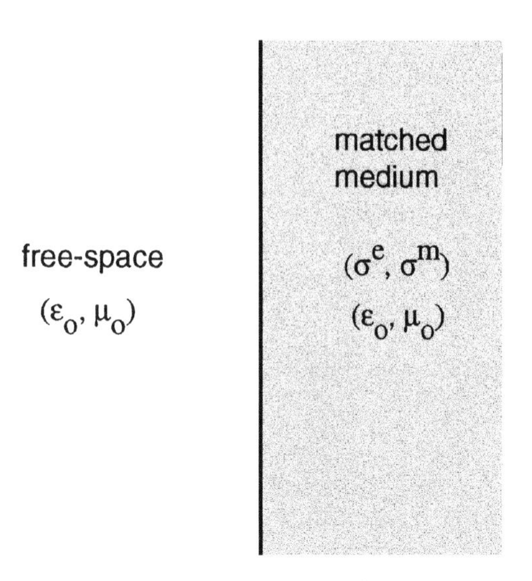

Figure 3.8 Interface Between Free-Space and a Perfectly Matched Layer Material

$$\varepsilon_0 \frac{\partial E_x}{\partial t} + \sigma^e E_x = \frac{\partial H_z}{\partial y} \tag{3.48}$$

$$\varepsilon_0 \frac{\partial E_y}{\partial t} + \sigma^e E_y = \frac{-\partial H_z}{\partial x} \tag{3.49}$$

$$\mu_0 \frac{\partial H_z}{\partial t} + \sigma^m H_z = \frac{\partial E_x}{\partial y} - \frac{\partial E_y}{\partial x} \tag{3.50}$$

For a plane wave traveling normal to the interface, perfect absorption takes place if the following condition is satisfied:

$$\frac{\sigma^e}{\varepsilon_0} = \frac{\sigma^m}{\mu_0} \tag{3.51}$$

In a typical FDTD simulation of radiating devices, waves do not travel in a uniform direction, thus limiting the utility of the matched layer.

The Perfectly Matched Layer (PML) was developed as an extension of the matched layer. In the PML, the absorption can be made possible for waves traveling at all angles and for any frequency. For the two-dimensional TE polarization case, the PML medium is constructed by splitting the H_z field into two non-physical components H_{zx} and H_{zy}, and assigning a respective electric and magnetic conductivity to each of the new field components. The new equations are:

$$\varepsilon_0 \frac{\partial E_x}{\partial t} + \sigma_y^e E_x = \frac{\partial (H_{zx} + H_{zy})}{\partial y} \tag{3.52}$$

$$\varepsilon_0 \frac{\partial E_y}{\partial t} + \sigma_x^e E_y = \frac{-\partial (H_{zx} + H_{zy})}{\partial x} \tag{3.53}$$

$$\mu_0 \frac{\partial H_{zx}}{\partial t} + \sigma_x^m H_{zx} = -\frac{\partial E_y}{\partial x} \tag{3.54}$$

$$\mu_0 \frac{\partial H_{zy}}{\partial t} + \sigma_y^m H_{zy} = \frac{\partial E_x}{\partial y} \tag{3.55}$$

where σ_x^e, σ_y^e are electric conductivities, and σ_x^m, σ_y^m are magnetic conductivities. Such PML medium is characterized by the parameters (σ_x^e, σ_y^e, σ_x^m, σ_y^m). To obtain the condition for perfect absorption across a free-space PML medium, with the interface parallel to the y-axis, we set

$$(\sigma_x^e, \sigma_x^m, \sigma_y^e, \sigma_y^m) = (\sigma_x^e, \sigma_x^m, 0, 0) \tag{3.56}$$

with

$$\frac{\sigma_x^e}{\varepsilon_0} = \frac{\sigma_x^m}{\mu_0} \tag{3.57}$$

For an interface parallel to the x-axis, we set:

$$(\sigma_x^e, \sigma_y^e, \sigma_x^m, \sigma_y^m) = (0, 0, \sigma_y^e, \sigma_y^m) \tag{3.58}$$

with

$$\frac{\sigma_y^e}{\varepsilon_0} = \frac{\sigma_y^m}{\mu_0} \tag{3.59}$$

For TM polarization, the E_z field is split in analogous manner to the TM case giving rise to E_{zx} and E_{zy}. For TM fields, Maxwell's equations take the following form

$$\varepsilon_0 \frac{\partial E_{zx}}{\partial t} + \sigma_x^e E_{zx} = \frac{\partial H_y}{\partial x} \tag{3.60}$$

$$\varepsilon_0 \frac{\partial E_{zy}}{\partial t} + \sigma_y^e E_{zy} = \frac{-\partial H_x}{\partial y} \tag{3.61}$$

$$\mu_0 \frac{\partial H_x}{\partial t} + \sigma_y^m H_x = -\frac{\partial (E_{zx} + E_{zy})}{\partial y} \tag{3.62}$$

$$\mu_0 \frac{\partial H_y}{\partial t} + \sigma_x^m H_y = \frac{\partial (E_{zx} + E_{zy})}{\partial x} \tag{3.63}$$

The same perfect matching conditions given in (3.61) and (3.62) apply equally to the TM polarization case.

The extension to three-dimensional space is analogous to the development above. Here, all six field components are split, resulting in the following 12 equations:

$$\mu_0 \frac{\partial H_{xy}}{\partial t} + \sigma_y^m H_{xy} = -\frac{\partial(E_{zx} + E_{zy})}{\partial y} \tag{3.64}$$

$$\mu_0 \frac{\partial H_{xz}}{\partial t} + \sigma_z^m H_{xz} = \frac{\partial(E_{yx} + E_{yz})}{\partial z} \tag{3.65}$$

$$\mu_0 \frac{\partial H_{yz}}{\partial t} + \sigma_z^m H_{yz} = -\frac{\partial(E_{xy} + E_{xz})}{\partial z} \tag{3.66}$$

$$\mu_0 \frac{\partial H_{yx}}{\partial t} + \sigma_x^m H_{yx} = \frac{\partial(E_{zx} + E_{zy})}{\partial x} \tag{3.67}$$

$$\mu_0 \frac{\partial H_{zx}}{\partial t} + \sigma_x^m H_{zx} = -\frac{\partial(E_{yx} + E_{yz})}{\partial x} \tag{3.68}$$

$$\mu_0 \frac{\partial H_{zy}}{\partial t} + \sigma_y^m H_{zy} = \frac{\partial(E_{xy} + E_{xz})}{\partial y} \tag{3.69}$$

$$\varepsilon_0 \frac{\partial E_{xy}}{\partial t} + \sigma_y^e E_{xy} = \frac{\partial(H_{zx} + H_{zy})}{\partial y} \tag{3.70}$$

$$\varepsilon_0 \frac{\partial E_{xz}}{\partial t} + \sigma_z^e E_{xz} = -\frac{\partial(H_{yx} + H_{yz})}{\partial z} \tag{3.71}$$

$$\varepsilon_0 \frac{\partial E_{yz}}{\partial t} + \sigma_z^e E_{yz} = \frac{\partial(H_{xy} + H_{xz})}{\partial z} \tag{3.72}$$

$$\varepsilon_0 \frac{\partial E_{yx}}{\partial t} + \sigma_x^e E_{yx} = -\frac{\partial(H_{zx} + H_{zy})}{\partial x} \tag{3.73}$$

$$\varepsilon_0 \frac{\partial E_{zx}}{\partial t} + \sigma_x^e E_{zx} = \frac{\partial(H_{yx} + H_{yz})}{\partial x} \tag{3.74}$$

$$\varepsilon_0 \frac{\partial E_{zy}}{\partial t} + \sigma_y^e E_{zy} = -\frac{\partial(H_{xy} + H_{xz})}{\partial y} \tag{3.75}$$

Theoretically, the PML provides a perfect absorption for traveling waves having any angle of incidence. However, in practice, the PML space has to be terminated. This typically involves the use of a perfect electric conductor, which introduces a reflection back into the computational space. A PML is therefore characterized by three parameters: thickness, conductivity profile, and the theoretical reflection coefficient at normal incidence. Clearly, the thicker the layer the higher the absorption. Empirical studies have shown that an effective conductivity profile is given by

$$\sigma(x) = \sigma_{max}(x/W)^3 \tag{3.76}$$

where W is the width of the PML. The corresponding theoretical reflection at normal incidence is given by

$$R(0) = e^{-\frac{\sigma_{max}W}{2\varepsilon_0 c}} \tag{3.77}$$

3.7 Field Extension

Most EMI/EMC problems involve an open region in which the domain of the solution covers the entire space. Open region electromagnetics radiation problems span wide disciplines within EMC applications. A classification of these problems can be made according to the location of the physically measurable quantity of interest which can be located at three different zones with respect to the structure under study: (1) very close to the object under study, such as when calculating the S parameters of a microstrip transmission line; (2) in the near zone region of the object as encountered in electromagnetic compatibility and interference studies (EMC/EMI); or (3) at an electrically and physically large distance from the object as the case would be when studying the emissions from computers enclosures to meet regulatory standards.

The FDTD method can be a powerful tool to solve open region problems only when applied in conjunction with a good mesh truncation technique. However, depending on the particular type of problem, the efficiency of the FDTD code can be variant as to render the method at times impractical. More specifically, when the objective is to determine the fields in the near zone region, enlarging the FDTD computational domain to enclose the near zone observation points can be very costly in terms of computer resources and run time. In fact, for many practical problems, the desired frequency band of the simulation and the fine resolution of the structure make direct FDTD calculation of the near zone simply impossible. However, even if resources were abundant, and inefficiency is tolerated, brute force direct FDTD calculation is not recommended for three reasons:

1. When enlarging the FDTD computational domain to include the near zone points, the accuracy of the ABC is jeopardized, since it experiences higher levels of energy incident at oblique angles, which in general will be poorly absorbed.

2. Grid dispersion errors increase with the size of the computational domain. These errors can be detrimental in applications requiring high phase accuracy.
3. The larger the domain, the longer the time needed for the simulation and hence the potential for instability in the time-marching scheme increases.

Therefore, instead of direct FDTD computation, Field Extension (FE) techniques are used to calculate the field outside the FDTD computational domain. These techniques invoke Huygens principle which requires finding the magnetic **M** and electric **J** currents on an imaginary surface that fully encloses the primary and secondary sources of radiation. There are several FE techniques available for calculating the field outside the FDTD computational domain, and all these depend on Huygen's principle. A convenient formulation is the Kirchhoff's Surface Integral Representation (KSIR). Only the KSIR formulation will be offered here, as it offers several advantages:

1. The KSIR formulation does not involve time integration terms, and thus can be much more efficient to implement.
2. To calculate any of the six field components at any location outside the surface, one need only to integrate an expression involving that same field component over the closed surface. This eliminates field interpolation that is essential to obtain **M** and **J** on the same surface.
3. The KSIR allows for calculating all the field components independently. This allows for direct and simple parallelization of the FE code.

Here, the KSIR formulation is presented as it applies to a Yee-cell based FDTD algorithm.

The Kirchhoff's integral representation is a relationship between the field inside a closed volume V and the field and its derivatives on the surface of V:

$$\psi(\mathbf{x},t) = \frac{1}{4\pi}\oint_{S'} \hat{n}\cdot\left\{\frac{\nabla'\psi(\mathbf{x}',t')}{R} - \frac{\mathbf{R}}{R^3}\psi(\mathbf{x}',t') - \frac{\mathbf{R}}{cR^2}\frac{\partial\psi(\mathbf{x}',t')}{\partial t'}\right\}_{ret} dS' \tag{3.78}$$

where $\mathbf{R} = \mathbf{x} - \mathbf{x}'$, $R = |\mathbf{R}|$, $\hat{n}$ is the unit normal vector to the surface, and c is the speed of light in free space; *ret* indicates that the integrand is evaluated at the retarded time $t' = t - R/c$.

The KSIR was originally derived as an approximation tool for predicting radiation from apertures. In the original development, the field distribution over the aperture or surface is approximated as the incident or primary field (typically the known field in the absence of the aperture). In FDTD the KSIR is used as a precise expression for determining the field in the exterior of the computational domain from the fields on a surface that completely encloses the radiating objects, as shown in Figure 3.9.

The standard Yee-cell-based implementation is well suited for the KSIR, since ψ can be any of the six field components. This feature allows for calculation of each component independent of the others.

Here, we outline the steps needed for simple implementation of (3.78) into a standard FDTD code based on the Yee cell. To illustrate

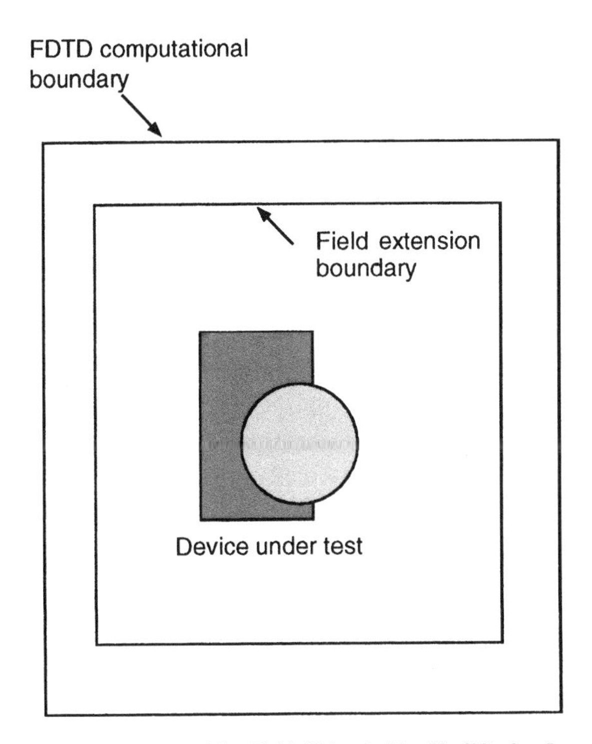

Figure 3.9 Calculations of Near- and Far-Fields Using the Korchhoff Surface Integral Representation Method

how (3.78) is discretized for FDTD implementation, we consider the value of the integral over only one of the six plane surfaces. Without loss of generality, we choose the surface in the x,y-plane, at $k = k_o$, where k is the FDTD z-index and, and let $\psi = E_x$. We then employ a change of variable in the time domain; that is instead of using retarded time, we express the calculated field as a function of the advanced time $t^* = t + R/c$. We denote the contribution to the integral over this surface as $\mathbf{E}_{x,k_o}$. At time step $n + 1$, the time and z-derivatives are approximately by a second-order accurate center difference formula:

$$\left.\frac{\partial E_x}{\partial z}\right|_{z=k_o} \cong D_z E_x^{n+1}(i,j,k_o) = \frac{E_x^{n+1}(i, j, k_o + 1) - E_x^{n+1}(i, j, k_o - 1)}{2\Delta z} \tag{3.79}$$

$$\left.\frac{\partial E_x}{\partial t}\right|_{t=n+1} \cong \frac{E_x^{n+2}(i, j, k_o) - E_x^{n}(i, j, k_o)}{2\Delta t} \tag{3.80}$$

For each observation point, R is a continuous function whose domain is the integration surface. However, in numerical evaluation of the integral where R is defined with respect to each cell, R becomes a discrete function with nonuniform spacing, and thus $E_x(\mathbf{x}, t^*)$ takes on values at time instances not separated by the FDTD time step Δt. Therefore, for uniformity and ease of implementation, we present $\mathbf{E}_x$ as a discrete sequence with a uniform time step equivalent to Δt. The time sequence index corresponding to t^* is denoted by n^* and is equal to the nearest integer to $|(n + 1) + R/(c\Delta t)|$, which we define as $n^* = |(n + 1) + R/(c\Delta t)|_{int}$. Substituting (3.78) and (3.79) in (3.78), and using staircase approximation of the integrand in (3.78), we have

$$E_{x,k_o}(\mathbf{x},t_n^*) = \sum_{i'j'} \{A\, D_z E_x^{n+1}(i',j',k_o) + B\, E_x^{n+1}(i',j',k_o) + \frac{C}{2\Delta t}\,[E_x^{n+2}(i',j',k_o) - E_x^{n}(i',j',k_o)]\}\, \Delta_{i'j'} \tag{3.81}$$

where

$$A = \frac{1}{4\pi}\frac{1}{R}$$

$$B = \frac{1}{4\pi}\frac{-\cos\theta'}{R^3}$$

$$C = \frac{1}{4\pi}\frac{-\cos\theta'}{cR^2}$$

In (3.81) **R** is vector pointing from each subsurface $\Delta_{i'j'}$ to the observation point x, and θ' is the angle that R' makes with the normal vector to the subsurface $\Delta_{i'j'}$. The primed indices $i'j'$ refer to the summation surface and should be differentiated from the usual FDTD indices i, j.

Finally, combining and sorting the three terms in (3.81) according to their time arguments, we have:

$$E_{x,k_o}(x,t_n^*) = F_1(n) + F_2(n+1) + F_3(n+2) \tag{3.82}$$

where

$$\begin{aligned} F_1(n) &= \sum_{i'j'} \frac{-C}{2\Delta t} E_x^n(i',j',k_o)\Delta_{i'j'} \\ F_2(n+1) &= \sum_{i'j'} (AD_z + B)\, E_x(i',j',k_o,n+1)\Delta_{i'j'} \\ F_3(n+2) &= \sum_{i'j'} \frac{C}{2\Delta t} E_x(i',j',k_o n+2)\Delta_{i'j'} \end{aligned} \tag{3.83}$$

Notice that no additional storage requirements are needed that are typically associated with FDTD time-derivative implementation. The appropriate contribution to E_x is made as the FDTD loop is executed. To illustrate this, we note that E_x is a sequence in discrete time. At the nth time step, $F_1(n)$, $F_2(n)$ and $F_3(n)$ are computed and only $F_1(n)$ is added to the register $E_x(n^*)$ (remember that $n^* = |(n+1) + R'/(c\Delta t)|_{int}$). In the next iteration, $F_1(n+1)$, $F_2(n+1)$ and $F_3(n+1)$ are calculated and only $F_2(n+1)$ contributes to $E_x(n^*)$. The final contribution to $E_x(n^*)$ comes from $F_3(n+2)$ which is calculated at the time step $n+2$. In each iteration, the F_i terms that do not contribute to $E_x(n^*)$ are added to either of the two registers $E_x(n^*-1)$ or $E_x(n^*+1)$ in a consistent manner. This procedure is repeated for each subsurface $\Delta_{i'j'}$. This scheme is explained in the diagram shown in Figure 3.10, where the arrows indicate the contribution to the E_x sequence from the FDTD iterations. Finally, the contribution of the remaining five surfaces is added in a similar manner.

The added cost to implementing the KSIR is due to the multiplications and additions required to calculate the F_i terms. The only increase in

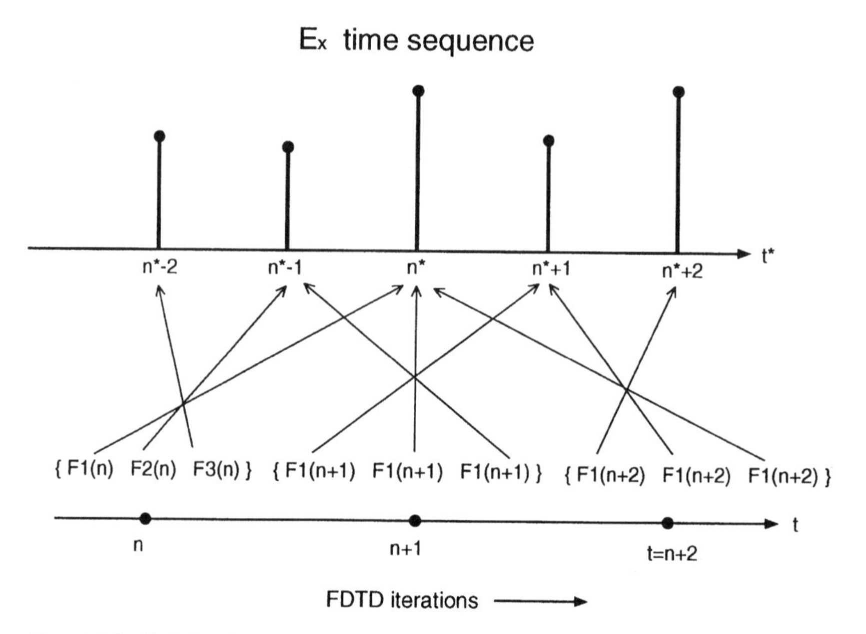

Figure 3.10 Updating the Fields Using the Korchhoff Surface Integral Representation Method

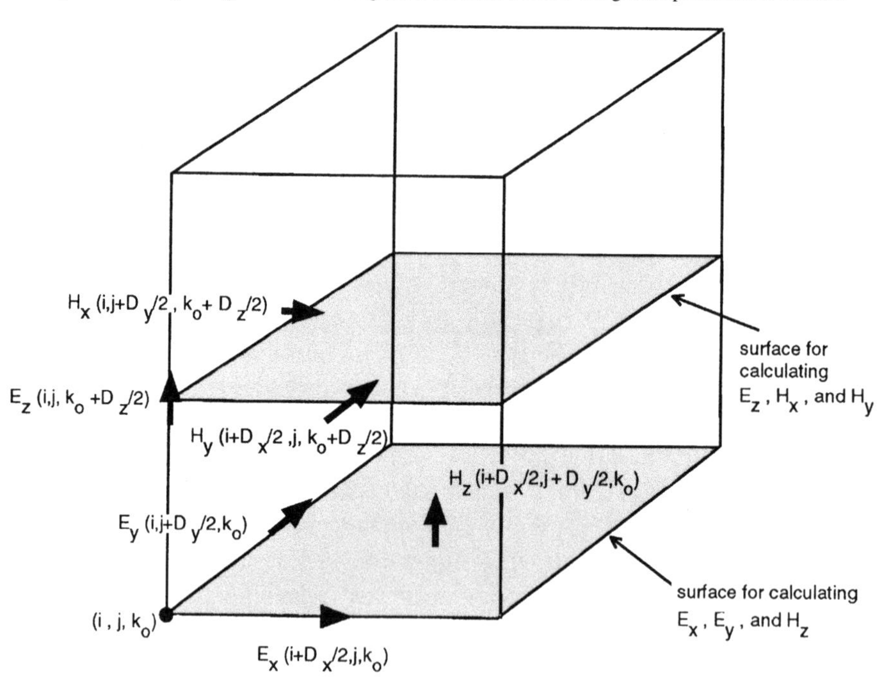

Figure 3.11 Yee Cell Showing Integration Surfaces

memory requirement comes from the storage of the two parameters R and θ', which are calculated once for each of the subsurfaces $\Delta_{i'j'}$. Therefore, the maximum increase in computer memory is $O(N^2)$, where N is the largest dimension of the computational box. As will be discussed in the section on validation, optimal results are obtained for the surface providing the tightest fit to the structure.

The remaining five field components are calculated in a similar manner. In standard Yee cell FDTD implementation, the **E** and **H** field components are one half-cell apart. E_x, E_y, and H_z all lie on one surface in the $x - y$ plane as shown in Figure 3.11, and therefore, these three fields can be calculated using the same surface used above. The remaining three field components, E_z, H_x, and H_y, lie on a second surface which is $\Delta_z/2$ apart from the first surface. The difference in the location of the two surfaces, however, does not introduce any complications since each field is calculated independently from the rest, and therefore a unique surface can be used for each field component. The only requirement dictated by KSIR is that when calculating each field component, the surface used must completely enclose the sources of radiation.

3.8 FDTD Simulation Errors

As has been outlined thus far, the calculation of the fields using the FDTD method involves several approximations. The most apparent and systematic approximations relate to the transformation of Maxwell's equations from differential equations to difference equations. Such approximation constitute a controled environment in the sense that the resulting errors can be decreased by creating a finer mesh. The primary errors that fall intor this category are the differencing errors arising from the staggering of **E** and **H** one half-space cell apart, and the grid anisotropy and dispersion.

A second source of error is due to the mesh truncation technique. This type of error can be made smaller, however, at the expense of increased memory overhead and operation count. The third source of error arise from inexact modeling of the physical source that couples energy into the system. This source of error can be one of the most challenging for EMI/EMC modelers. A diagram illustrating these types of FDTD errors is given in Figure 3.12.

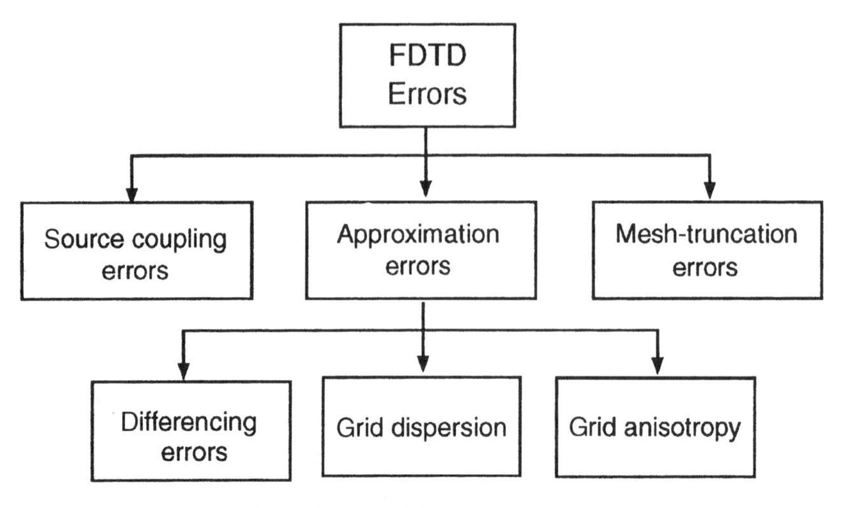

Figure 3.12 Modeling Errors in FDTD

References

1. S.D. Conte and C. de Boor, *Elementary Numerical Analysis: An Algorithmic Approach*, McGraw-Hill, New York, 1980.
2. K.E. Atkinson, *An Introduction to Numerical Analysis*, John Wiley & Sons, New York, 1978.
3. K.S. Yee, "Numerical solution of initial value problems involving Maxwell's equations in isotropic media," *IEEE Transactions on Antennas and Propagation*, vol. 14, pp. 302–307, 1966.
4. A. Taflove, *Computational Electrodynamics: The Finite-Difference Time-Domain Method*, Artech House: Boston, 1995.
5. K.S. Kunz and R.J. Luebbers, *The Finite Difference Time Domain Method for Electromagnetics*, CRC Press, Boca Raton, FL, 1993.
6. R.L. Higdon, "Radiation Boundary Conditions for Elastic Wave Propagation," *SIAM Journal of Numerical Analysis*, vol. 27, pp. 831–870, 1990.
7. O.M. Ramahi, "Complementary operators: A method to annihilate artificial reflections arising from the truncation of the computational domain in the solution of partial differential equations," *IEEE Transactions on Antennas and Propagation*, vol. 43, pp. 697–704, 1995.
8. O.M. Ramahi, "Complementary boundary operators for wave propagation problems," *Journal of Computational Physics*, vol. 133, pp. 113–128, 1997.
9. J-P. Berenger, "A perfectly matched layer for the absorption of electromagnetic waves," *Journal of Computational Physics*, vol. 114, pp. 185–200, 1994.

10. J-P. Berenger, "Three-dimensional perfectly matched layer for the absorption of electromagnetic waves," *Journal of Computational Physics*, vol. 127, pp. 363–379, 1996.
11. R.L. Higdon, "Absorbing boundary conditions for acoustic and elastic waves in stratified media," *Journal of Computational Physics*, vol. 101, pp. 386–418, 1992.
12. R.L. Higdon, "Radiation Boundary Conditions for Dispersive Waves," *SIAM Journal of Numerical Analysis*, vol. 31, pp. 64–100, 1994.
13. M.J. Barth and R.R. McLeod and R.W. Ziolkowski, "A near and far-field projection algorithm for Finite-Difference Time-Domain codes," *Journal of Electromagnetic Waves and Applications*, vol. 6, pp. 5–18, 1992.
14. I.J. Craddock and C.J. Railton, "Application of the FDTD method and a full time-domain near-field transform to the problem of radiation from a PCB," *Electronic Letters*, vol. 29, pp. 2017–2018, 1993.
15. R.J. Luebbers, K.S. Kunz, M. Schneider, and F. Hunsberger, "A finite-difference time-domain near zone to far zone transformation," *IEEE Transactions on Antennas and Propagation*, vol. 39, pp. 429–433, 1991.
16. C.L. Britt, "Solution of electromagnetic scattering problems using time domain techniques," *IEEE Transactions on Antennas and Propagation*, vol. 37, pp. 1181–1192, 1989.
17. J.D. Jackson, *Classical Electrodynamics*, John Wiley & Sons, New York, 1962.

Chapter 4

Method of Moments

4.1 Introduction

The Method of Moments (MoM) is part of a general body of mathematical techniques whose goal is to solve an integral equation by converting it into a matrix equation, which can then be readily solved on a computer. The MoM is one of the most well-developed and used of all numerical techniques available in electromagnetic analysis, including EMI/EMC work. The MoM is very versatile; it can be highly efficient for metallic structures composed of wires and surfaces.

The MoM differs markedly from volume-based methods such as the Finite-Difference Time-Domain (FDTD) and Finite Element method (FEM), in that it is a surface-based method. By this we mean that only the body or physical structure of the object being analyzed is discretized, or converted into discrete entities, and then fed into computer memory for analysis. This makes the MoM highly efficient when treating the type of problem having perfectly conducting wire scatterers without the presence of any electromagnetically penetrable bodies, such as dielectric and magnetic materials. Furthermore, since the MoM is a frequency-domain technique, it can be a very efficient analysis tool when the response of the structure is desired over a single frequency or very narrow band of frequency. (The MoM can also be used to solve time-domain, or transient problems; however, the time-domain integral equations are much more difficult to analyze and are beyond the scope of this book.)

The MoM procedure comprises four steps: (1) dividing the structure to be modeled into a series of wire segments and patches whose dimensions are much shorter than the wavelength of interest; (2) choosing expansion functions to represent the unknown current, and the weighting functions; (3) filling the matrix elements and solving for the unknown current distribution on the body of the structure; and (4) postprocessing the output current values for near-field, far-field, or other desired characteristics of the system such as power and impedance.

The MoM can be used to treat not only conductive structures, but complex inhomogeneous media as well. As this chapter is intended to be introductory, the discussion will be limited to perfectly conducting bodies only.

4.2 Linear Operators

The MoM refers to a general procedure for solving linear mathematical equations of the form:

$$L(u) = g \tag{4.1}$$

where L is a linear operator, typically of the integral type, u is the unknown function we need to solve for, and g is a known excitation function that represents the primary source of energy in the system. A linear operator satisfies the following identity:

$$L(au_1 + bu_2) = aL(u_1) + bL(u_2) \tag{4.2}$$

where a and b are constants, and u_1 and u_2 are independent functions. The linearity of an operator will prove fundamental to the MoM formulation, as will be discussed.

In electromagnetic applications, the known excitation g is usually an impressed electric or magnetic field, or an impressed current density. The unknown function, u, is typically the current distribution on the structure. The MoM that can be solved by the MoM has the following form:

$$L(u) = \int uK = g \tag{4.3}$$

Equation (4.3) is referred to as an integral equation of the first kind, meaning that the unknown function u appears only under the integral. The function K in (4.3) is a predetermined function that depends on the space of the problem, but not on the radiating or scattering object itself.

Recall that Maxwell's equations are coupled partial differential equations. To solve for one of the two unknown fields in Maxwell's equations, the equations have to be manipulated to obtain a single integral equation of only one of the two unknown fields. Because of the direct interdependence of the electric and magnetic fields, as specified by Maxwell's equations, obtaining one of the two fields gives a complete solution of the problem.

Maxwell's equations can be cast into a variety of integral equations. Here, and for the purpose of demonstrating the MoM, the focus will be on one of the most popular integral equations describing the behavior of the **E** field, the Pocklington Integral Equation.

4.3 Pocklington Integral Equation

Several integral equations describe the relationship between an excitation source and the electric and magnetics currents on the radiating structure. Here, the MoM will be demonstrated by applying it to the Pocklington integral equation, one of the most common and versatile equations used to describe the behavior of radiating surfaces.

To derive the Pocklington equation, Consider a wire segment oriented in the z-direction, as shown in Figure 4.1. Assuming a current flowing along the wire surface (the wire is assumed to be a perfect conductor), the vector potential **A** (see Chapter 2) is readily expressed as:

$$\tilde{\mathbf{A}}(r) = \mu \iint_s \tilde{\mathbf{J}}(r') \frac{e^{-jkR}}{4\pi R} dr' \tag{4.4}$$

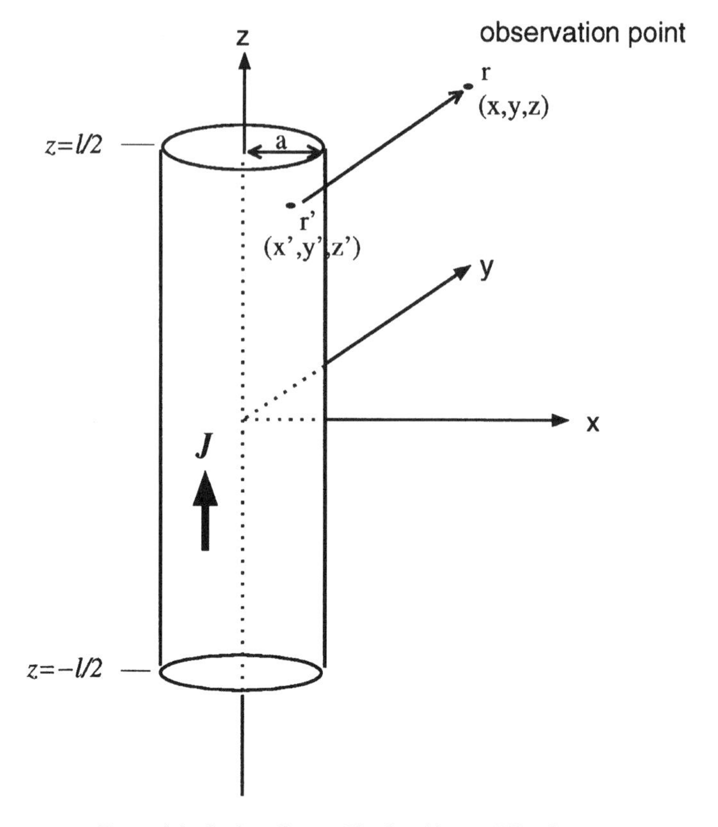

Figure 4.1 Surface Current Flowing Along a Wire Segment

where $\tilde{\mathbf{J}}$ is the surface current density, and R is the distance from the integration point on the surface of the structure to the observation point:

$$R = \sqrt{(x - x')^2 + (y - y')^2 + (z - z')^2} \tag{4.5}$$

The electric field at any point in space is obtained from $\tilde{\mathbf{A}}$ as:

$$\tilde{\mathbf{E}}(r) = -j\omega\tilde{\mathbf{A}} - j\frac{1}{\omega\mu\varepsilon}\nabla(\nabla \cdot \tilde{\mathbf{A}}) \tag{4.6}$$

Assuming that the current travels along the wire in the direction parallel to its axis, which is the z-direction in the wire segment shown in Figure 4.1. In such a case, only A_z exists and is given by:

$$A_z = \frac{\mu}{4\pi} \int_{-l/2}^{l/2} \int_0^{2\pi} J_z \frac{e^{-jkR}}{R} \, d\phi' \, dz' \tag{4.7}$$

And the resulting E_z field reduces to:

$$E_z = -j\frac{1}{\omega\mu\varepsilon}\left(k^2 A_z + \frac{\partial^2 A_z}{\partial z^2}\right) \tag{4.8}$$

After substituting (4.6) in (4.7), we have

$$E_z = \frac{-j}{4\pi\omega\varepsilon} \int_{-l/2}^{l/2} \int_0^{2\pi} \left(k^2 \frac{e^{-jkR}}{R} + \frac{\partial^2}{\partial z^2} \frac{e^{-jkR}}{R}\right) J_z \, dz' \tag{4.9}$$

Further simplification is obtained by assuming that the current distribution is uniform with respect to ϕ. This assumption is referred to as the thin-wire approximation and is typically used when the radius of the wire, a is less than 0.1 λ. Under this assumption, the current density, J_z no longer has azimuthal variation and is expressed as

$$J_z = \frac{1}{2\pi a} I_z(z') \tag{4.10}$$

Here, I_z is assumed to be an equivalent filament line current located on the wire surface. Further, simplification is made by observing the field at the wire axis instead of the surface of the wire giving $R = [a^2 + (z - z')^2]^{-\frac{1}{2}}$. Finally, (4.9) reduces to:

$$E_z = \frac{-j}{4\pi\omega\varepsilon} \int_{-l/2}^{l/2} \left(k^2 + \frac{\partial^2}{\partial z^2}\right) G(z,z') I_z \, dz' \tag{4.11}$$

where

$$G(z,z') = \frac{e^{-jkR}}{4\pi R} \tag{4.12}$$

The field, E_z given by (4.12) is the result of the current I_z flowing along the wire segment. On the surface of the wire (reduced for the purpose of simplification to the axis of the wire), the total tangential electric field is zero, since the wire is assumed to be perfectly conducting. The total field on the wire surface is the sum of the field due to the current I_z and the incident field (or impressed field) E_z^{inc}, which is a known quantity. Therefore, we have:

$$E_z = -E_z^{inc} \tag{4.13}$$

Finally substituting (4.13) into (4.11), we obtain:

$$E_z^{inc} = \frac{j}{2\pi\omega\varepsilon} \int_{-l/2}^{l/2} \left(k^2 + \frac{\partial^2}{\partial z^2}\right) G(z,z')\, I_z dz' \tag{4.14}$$

Equation (4.14) is referred to as the Pocklington integral equation. It describes the variation of the current along the surface of the wire, given an incident electric field.

By observation of (4.14) and (4.1), the integral operator defining the Pocklington integral equation is given by

$$L(E) = \frac{j}{2\pi\omega\varepsilon} \int_{-l/2}^{l/2} \left(k^2 + \frac{\partial^2}{\partial z^2}\right) G(z,z')\, E dz' \tag{4.15}$$

and the forcing function g in (4.1) is now E_z^{inc}.

4.4 Method of Moments Development

Once the linear operator relating the unknown current to the impressed field is obtained, the goal of the MoM is to solve the operator equation by transforming it into a matrix equation. The MoM technique requires the creation of a system of N linear equations having N unknowns, where each unknown represents the current magnitude on a single wire segment or surface patch. For simplicity, we will only be concerned with wire structures, since the inclusion of surface patches requires advanced analysis beyond the scope of this introductory chapter.

4.4.1 Matrix Construction

The first step in the MoM solution process is to describe the unknown current distribution over a segmented wire structure (see Figure 4.2)

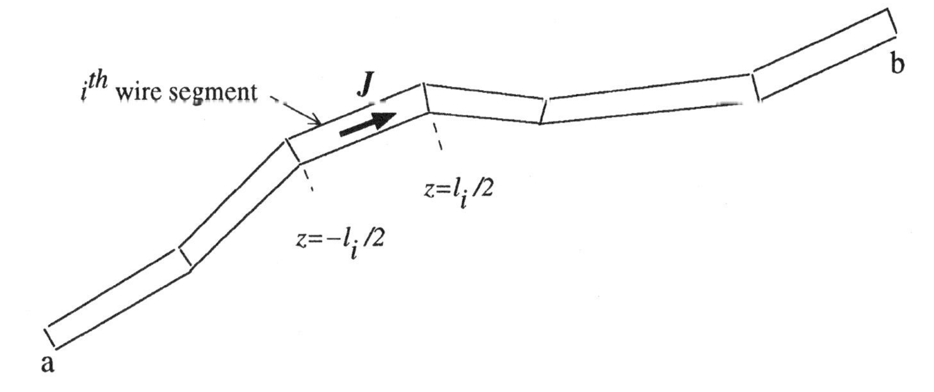

Figure 4.2 Curved Wire Approximated as a Series of Short Wire Segments

as a linear combination of functions with unknown multiplicative coefficients as:

$$I(z) = \sum_{i=1}^{N} I_i \; \psi_i(z) \tag{4.16}$$

Here, each ψ_i' is a function intended to approximate the behavior of the current over the ith wire segment, and I_i is an unknown multiplicative coefficient. The remaining procedure of MoM is the determination of the best-suited I_i's that give the best approximation to the current distribution according to certain criterion. Define a Residual Error (RE) as:

$$RE = E_z^{inc} - L(I) \tag{4.17}$$

If the solution to the equation $L(I) = E_z^{inc}$ is exact, the residual error will be zero. The major theme of the MoM is to obtain a solution to $L(I) = E_z^{inc}$ by forcing the residual error to be zero, using an averaging or weighting process. In different words, a solution is sought that satisfies (4.17) in some statistical average. To do this, multiply both sides of (4.17) by a set of weighting or testing functions w_j, for j = 1,2, . . ., N, and integrate the product over the entire length of the wire. This gives:

$$\int_a^b RE \; w_j \; dl = \int_a^b \left[E_z^{inc} - L(I)\right] w_j \; dl = 0; \quad j = 1,2, \ldots, N \tag{4.18}$$

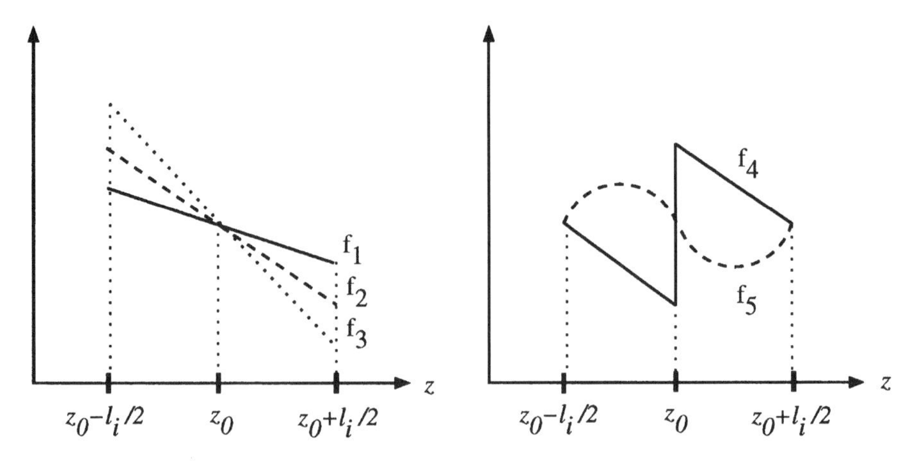

Figure 4.3 Curved Wire Approximated as a Series of Short Wire Segments

where a and b denote the end points of the wire. Notice that the averaging criterion imposed here is not very restrictive. It only demands that the average of a new function defined as the weighted residual error, $[E_z^{inc} - L(I)]\, w_j$ be zero. To give a sense of what is meant by this average, Figure 4.3 shows five different functions, f_1 to f_5. Notice that all these functions are distinctly different from each other, however, their average; i.e., their integral over the length l is equivalent.

Using the linearity of L and making the substitution (4.16) in (4.18), the right hand side of equation (4.18) becomes

$$\int_a^b [E_z^{inc} - L\,(\sum_{i=1}^{N} I_i\, \psi_i)]w_j\, dl = 0; \quad j = 1, 2, \ldots, N \qquad (4.19)$$

which reduces to

$$\sum_{i=1}^{N} I_i \int_a^b w_j\, L(\psi_i)\, dl = \int_a^b w_j\, E_z^{inc}\, dl; \quad j = 1, 2, \ldots, N \qquad (4.20)$$

To make the equations more concise, we set

$$Z_{ij} = \int_a^b L(\psi_i)\, w_j\, dl \qquad (4.21)$$

$$E_j = \int_a^b w_j\, E_z^{inc} \qquad (4.22)$$

Finally, substituting (4.21) and (4.22) into (4.20), we have:

$$
\begin{aligned}
Z_{11}I_1 + Z_{12}I_2 + \ldots + Z_{1N}I_N &= E_1 \\
Z_{21}I_1 + Z_{22}I_2 + \ldots + Z_{2N}I_N &= E_2 \\
&\;\;\vdots \\
Z_{N1}I_1 + Z_{N2}I_2 + \ldots + Z_{NN}I_N &= E_N
\end{aligned} \tag{4.23}
$$

which can be expressed in a matrix notation as:

$$
\begin{bmatrix}
Z_{11} & Z_{12} & \cdots & Z_{1N} \\
Z_{21} & Z_{22} & \cdots & Z_{2N} \\
\vdots & & & \\
Z_{N1} & Z_{N2} & \cdots & Z_{NN}
\end{bmatrix}
\begin{bmatrix} I_1 \\ I_2 \\ \vdots \\ I_N \end{bmatrix}
\begin{bmatrix} E_1 \\ E_2 \\ \vdots \\ E_N \end{bmatrix} \tag{4.24}
$$

or more concisely as:

$$[Z][I] = [E] \tag{4.25}$$

Notice that the final matrix system in (4.25) resembles Ohm's law, where $[Z]$ can be interpreted as a generalized impedance matrix, and the excitation $[E]$ can be interpreted as a generalized voltage matrix.

The cost of an MoM procedure is typically measured in terms of two parameters: The first is the number of segments that are needed to obtain a sufficiently accurate solution (solution convergence). The second is the time needed to fill the matrix elements Z_{ij}. The filling of the matrix $[Z]$ highly depends on the expansion functions used for the current and on the testing function, both to be discussed next.

4.4.2 Basis and Testing Functions

The preceding section outlined the general procedure employed in the MoM to convert or transform the analytic operator equation $L(I) = E_z^{inc}$ into a matrix equation that can be solved on computer. Clearly, the procedure outlined thus far leaves few questions unanswered, such as the choice of basis and testing functions essential for the completion of the solution. In theory, the class of functions that are admissible as basis and testing functions can be very large; however, practical and numerical considerations place a constraint on the functions that can be used and that can yield valid solutions. It is also important to keep in mind that the MoM is a numerical approximation procedure which does not guarantee the convergence of the solution independently of the choice of the basis and testing functions.

The simplest choice for current basis or expansion function is a series of pulse functions that collectively give a staircase approximation to the unknown current over the length of the wire. To see how these functions can be employed, the wire is divided into N segments. Over these segments, the pulse functions expansion is then given by:

$$I(z) = \sum_{i=1}^{N} I_i \, P_i(z) \tag{4.26}$$

where P_i is a constant function over the ith segment and zero over the remaining segments. This piecewise-constant approximation of a smoothly varying current distribution is shown in Figure 4.4. Notice that the segments need not be equal in length. When the current is expected (either from physical insight or previous modeling experience) to vary rapidly over a portion of the wire, smaller segments will be needed to capture this variation. On the other hand, if the current is expected to have a very slow variation over part of the wire, the segments can be enlarged to reduce the number of unknowns. However, in any segmentation employed, the length of a segment has to be much smaller than the wavelength.

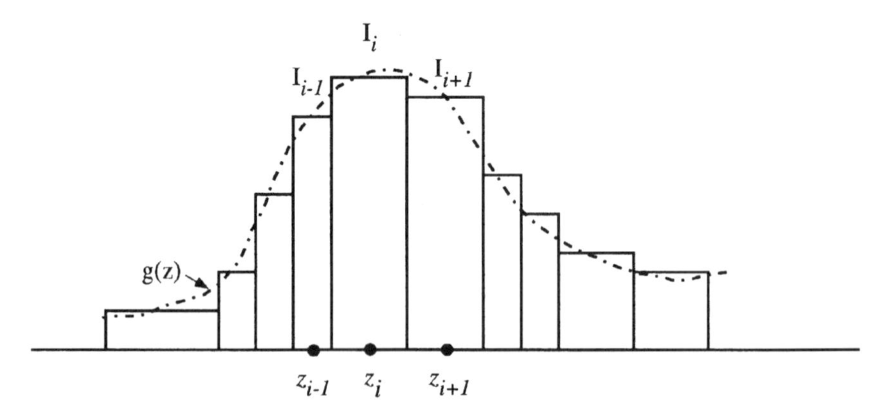

Figure 4.4 Staircase Approximation to the Function $g(z)$, Using Pulse Basis Functions

The next step to select the weighting (testing) functions. The simplest weighting functions are given by the Dirac-Delta functions $\delta\,(z' - z_n)$, where z_n denotes the center of the nth wire element. This choice is also referred to as *point-matching* or *collocation*, since it effectively sets the residual at each of the segment center points to zero. With the Dirac-Delta functions as the testing functions, the matrix element reduces to:

$$Z_{mn} = \int_{-\frac{l_i}{2}}^{\frac{l_i}{2}} G(z_n, z')\, dz' \tag{4.27}$$

The point-matching method generates the simplest possible matrix elements; however, the accuracy of the solution at points different from where the matching takes place generally cannot be guaranteed.

A more advanced choice for the testing functions would be to use pulse functions identical to the ones used for the expansion of the current along the wire. This choice results in the following matrix element:

$$Z_{ij} = \int_{-\frac{l_i}{2}}^{\frac{l_i}{2}} \int_{-\frac{l_j}{2}}^{\frac{l_j}{2}} G\,(z_n, z')\, dz' \tag{4.28}$$

When the current expansion functions are identical to the weighting functions, the procedure is referred to as the *Galerkin method*, which is probably the most popular method.

A more complex, but highly efficient current basis function is given by the piecewise-sinusoid function defined as

$$f(z') = \frac{\sin(z - z_{n-1})k}{\sin(z_n - z_{n-1})k}; \quad z_{n-1} < z' < z_n \tag{4.29}$$

$$f(z') = \frac{\sin(z_{n+1} - z)k}{\sin(z_{n+1} - z_n)k}; \quad z_n < z' < z_{n+1} \tag{4.30}$$

In (4.28), unlike the pulse basis functions, the current over each wire segment is described by two sinusoid functions. A Galerkin procedure employing the piecewise-sinusoid functions is found to require approximately 10 times fewer segments than would be the needed had pulse expansion functions been used.

One of the most popular and widely available MoM codes is the

Numerical Electromagnetic Code (NEC) developed by the Lawrence Livermore National Laboratory. NEC uses the point matching procedure (Dirac-Delta functions as the testing functions) and a different basis functions from the ones used above. The current expansion function over each segment used by NEC is given by:

$$I_i(z) = \alpha_i + \beta_i \sin(z - z_i)k + \gamma_i \cos(z - z_i) \tag{4.31}$$

The three coefficients α_i, β_i, and γ_i are related such that the continuity of current and charge is satisfied across segment junctions and ends. In contrast to the current expansion functions used earlier, equation (4.29) involves three unknowns resulting in an increased overhead. This can be considered the tradeoff for using more accurate basis functions with a faster convergence rate.

4.4.3 Matrix Solution

Once the matrix elements are computed and the system matrix (4.25) is fully determined, the current matrix $[I]$ is obtained by first inverting the matrix $[Z]$ and then solving the matrix equation

$$[I] = [Z]^{-1}[E] \tag{4.32}$$

The matrix equation (4.32) can be solved in a variety of methods with varying degrees of efficiency, depending on the structure of the impedance matrix $[Z]$. The symmetry in certain wire structures can be exploited to reduce the matrix fill time, as well as the matrix solution time. However, in general, the matrix $[Z]$ can be assumed full. For full matrices, an efficient solution is obtained using the Gauss-Jordan algorithm which gives a solution time directly proportional to N^2, where N is the number of unknowns.

With the current distribution along the wire segments known, the field at any point in space can be determined using (4.4) and (4.6). If only the far field is desired, the formula derived in Chapter 2 can be directly applied to the current.

This chapter presents the MoM as applied to perfectly conducting wire structures. If surface patches are part of the structure to be modeled, as a first model approximation, the patches are converted into a wire

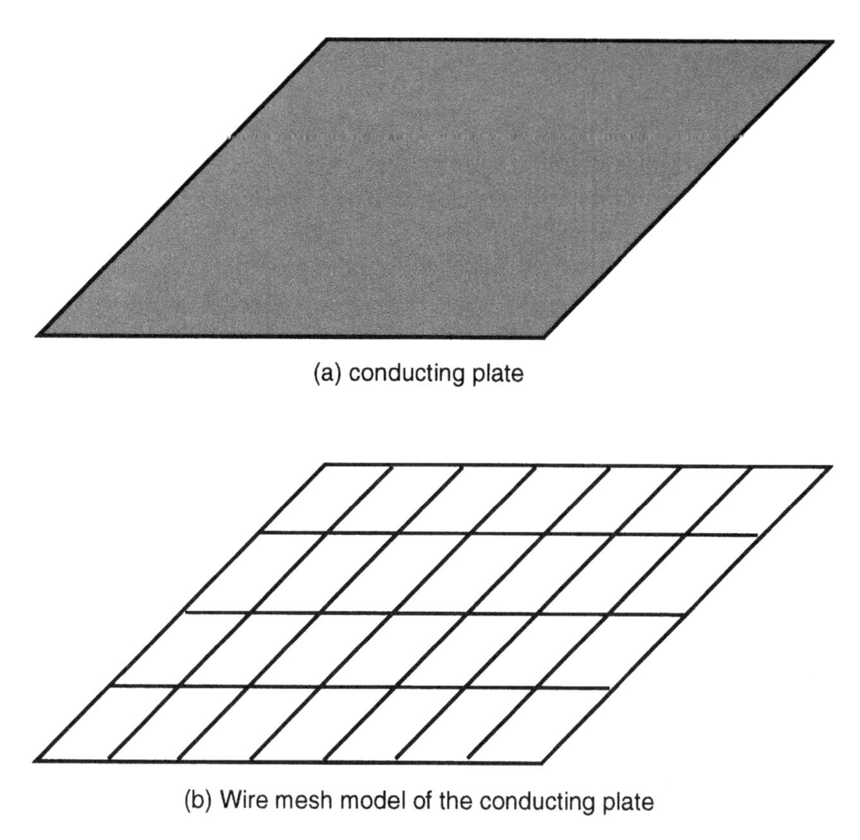

Figure 4.5 Approximating a Conducting Surface by Wire Grid: (a) Conducting surface. (b) Wire grid model approximating the conducting surface.

grid, as shown in Figure 4.5. However, it is possible to use the MoM to model surfaces by segmenting them into surface patches and expanding the current using two-dimensional basis functions in a manner that is fully analogous, yet more cumbersome, to the one-dimensional basis functions used for wire modeling. If surface patches are used, a different integral equation is used that employs the magnetic field instead of the electric field.

For many practical modeling problems, approximating the surface as a grid of wires can yield satisfactory results, especially if the far field is the end objective of the analysis. It is important to keep in mind that the wire-grid model restricts the direction of the current in the model and might not be suitable in applications where accurate current direction can affect accuracy. Also it should be noted that only axial current along the wires was allowed when deriving the MoM equations for wire structures. This excludes any current variation in the circumferencial direction. For such an approximation to hold and give good accuracy, the radius of the wire has to be much smaller than the wavelength λ. As mentioned earlier, the thin-wire approximation is considered valid when a is less than 0.1 λ.

Another factor that affects the accuracy of the MoM simulation is the ratio of the wire radius to wire segment length a/l. One would intuitively expect that the smaller the segment, the greater the accuracy of the model. This is actually true, but to a certain degree. Numerical experiments have shown that good accuracy is obtained if the ratio a/l is kept less than 0.1. If a/l is increased beyond this point, currents near free wire ends can exhibit oscillatory behavior that can introduce large errors in the solution.

4.5 Summary

Because of its versatility and relative ease of programming, the MoM has been used extensively in electromagnetics analysis from prediction of radar cross-section of fighter aircrafts to EMI/EMC studies of radiating elements on a computer motherboard. A large body of literature has been written on the MoM, with many improvements added that are suitable for special applications and structures. Furthermore, extensions of the MoM have been developed to treat wire structures with a ratio a/l greater than 0.1 and also for structures residing above a conducting ground plane. The major constraint of MoM is that it is a frequency-domain analysis algorithm. This implies that to obtain the solution over a wide band of frequencies, several runs would be necessary, with each run requiring the solution of a matrix equation.

References

1. R.F. Harrington, *Field Computation by Moment Methods*, Robert E. Krieger, Malabar, FL, 1987.
2. W.L. Stutzman and G.A. Thiele, *Antenna Theory and Design*, John Wiley & Sons, New York, 1981.
3. R.C. Booton, Jr., *Computational Methods for Electromagnetics and Microwaves*, John Wiley & Sons, New York, 1992.

Chapter 5

The Finite Element Method

5.1 Introduction

The Finite Element Method (FEM) is a numerical technique used to solve partial differential equations by transforming them into a matrix equation. The primary feature of FEM is its ability to describe the geometry or the media of the problem being analyzed with great flexibility. This is because the discretization of the domain of the problem is performed using highly flexible nonuniform patches or elements that can easily describe complex structures.

Several partial differential equations are of interest to EMI/EMC work in particular and to electromagnetics researchers in general. Covering the FEM treatment of all these equations can be a very lengthy exercise and is beyond the scope of an introductory chapter as intended here. However, and fortunately, the FEM concepts apply consistently to all these partial differential equations, except for some variations specific to the physics of the problem. Since the FEM is a numerical method, it will be introduced and discussed by applying it to solve the Laplace equation describing the potential distribution. The Laplace equation is a natural choice for introducing the FEM because of its simplicity; thus, careful exposition of the fundamentals of the method can be made possible without being sidetracked by complex mathematical formulations. Next, the application of FEM to solving the Helmholz wave equation is addressed. Because the Helmholz equation arises in many electromagnetics radiation problems in open space, it requires careful attention when it is solved using the FEM, since the radiation condition must be included and made part of the whole problem.

The FEM bears a strong resemblance to the Method of Moments (MoM) in the sense that both methods convert either a differential or an integral equation into a matrix equation. However, and as a distinct variation from the MoM, the FEM is based on the physical principle of minimizing the energy of the system. This gives the FEM an appeal that extends across several engineering disciplines.

This chapter will introduce the underlying theme of the FEM by first defining some mathematical preliminaries that are essential to understanding the FEM theory. While this mathematical background is fundamental to FEM theory, it will be attempted with a minimum amount of rigor in order to only highlight the basic principles of the method without alienating the nonmathematically inclined reader. For those readers who like to probe further into this subject, a list of excellent FEM references is provided towards the end of this chapter.

5.2 Variational Forms

Consider a partial differential equation that describes the function u. This equation can be represented symbolically by operator notation:

$$L(u) = f \tag{5.1}$$

where f is the known forcing or excitation function, and L represents a partial differential equation operator that typically involves partial derivatives of the function $\dot{u}$ with respect to its independent variables. Next, the inner product of two functions g and h may be defined as:

$$<g,h> = \int_{\Omega} g\ h\ d\Omega \tag{5.2}$$

where Ω is the domain of g and h.

The purpose of defining the inner product as in (5.2) is to facilitate concise mathematical representation, as will be shown in the following discussion.

Next, the concept of the functional is introduced. The functional, denoted as F, is defined as a mapping that assigns a number to a function. This makes the functional a logical extension of the concept of the function, which assigns a number to another number. For instance, if $g(x)$ is a known function, then $F[g(x)]$ will be a unique number assigned to the function $g(x)$. (In mathematics jargon, this mapping is referred to as one-to-one.) The functional can be viewed as a function of functions.

Having defined what a functional is, the following functional F given by:

$$F(v) = <L(v),v> -2 <f,v> \tag{5.3}$$

can be shown to have a minimum at the function $v = u$, where u is the solution to the partial differential equation in (5.1). For the functional F to have a minimum at $v = u$, it is meant that the first derivative, or first variation vanishes. To show this, assume that F has a minimum at u, then for ε small, we have:

$$F(u) \leq F(u + \varepsilon v) = F(u) + 2\varepsilon[<L(u), v> - <f,v>] + \varepsilon^2 <Lv,v> \tag{5.4}$$

Since F has a minimum, the first variation, which is the coefficient of ε, must be equal to zero. This gives:

$$<L(u),v> = <f,v> \tag{5.5}$$

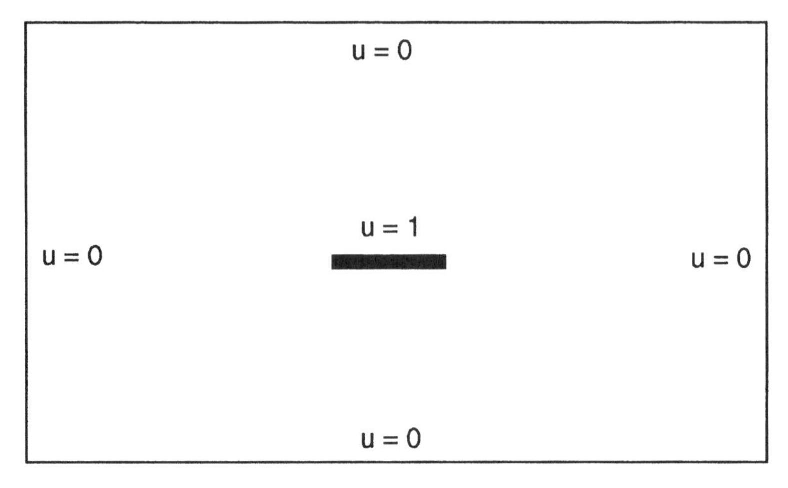

Figure 5.1 Configuration for a Shielded Microstrip Transmission Lines

For an arbitrary v, this implies

$$L(u) = f \tag{5.6}$$

Therefore, with the construction of F as in (5.3), the objective of solving the equation $L(u) = f$ becomes equivalent, in the mathematical sense, to minimizing the functional $F(u)$. The expression in (5.3) is called a variational form and it forms the foundation of the FEM.

The Laplace equation is considered one of the simplest partial differential equations encountered in electromagnetics theory. The Laplace equation arises in many applications where the interest lies in describing the potential distribution in structures such as transmission media. For instance, the potential distribution in the shielded microstrip line shown in Figure 5.1 can be determined by solving the Laplace equation with a given boundary conditions on the conductors and side walls. The simplicity of the Laplace equation makes it appealing to use as a good example to demonstrate the FEM theory.

The Laplace equation in two-dimensional space is given by:

$$\nabla^2 u(x,y) = 0 \tag{5.7}$$

Here, $u(x,y)$ represents the potential distribution. The excitation to this problem is indirectly provided by the boundary condition, which takes the form of u, and its normal derivative u_n being specified over part or all of the boundary of the domain. By analogy to (5.1), the differential operator L is given by

$$L = \nabla^2 \tag{5.8}$$

Following the recipe of (5.3), we construct a functional which achieves its minimum at a function that is the solution to (5.7). This functional is given by:

$$F(u) = \int_{\Omega} \nabla^2 u \; u \; d\Omega \tag{5.9}$$

where Ω is the domain where the potential is to be determined.

Using vector calculus identities, (5.9) can be transformed into the following form:

$$F(u) = \int_{\Omega} |\nabla u|^2 \; d\Omega \tag{5.10}$$

The functional obtained in (5.10) is linearly proportional to the stored energy in the system. In steady state, the system reaches its minimum energy. This minimum energy state is precisely the solution to the differential equation (5.9).

Notice that the functional or the integral in (5.10) has a domain which covers the entire space of the problem. This domain includes not only any radiating and scattering objects, but the empty space that surrounds them as well. This makes the FEM markedly different from the Method of Moments (MoM) (see Chapter 4), where the domain of the integral covers only the radiating objects and the scattering structures that surrounds it.

Once the variational form of the problem is fully constructed, the next step is to describe the field within the domain of the problem in terms of basis functions defined over patches or segments of the domain. These segments are called finite elements.

5.3 Construction of Finite Elements

The construction of the variational form is essential to the FEM solution of a partial differential equation; however, it is not an integral part of

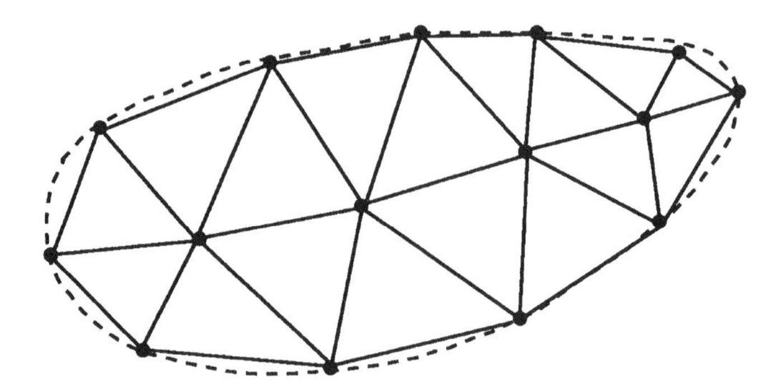

Figure 5.2 Approximation by Nonoverlapping Finite Elements

the FEM procedure. Strictly speaking, the FEM is a procedure designed to find the function that renders the functional minimum.

The basic concept of the FEM is to divide the domain or region of the problem into small connected patches, called finite elements. These finite elements are connected in a non-overlapping manner, in such a way that they completely cover the entire space of the problem. In two-dimensional space, the elements can take different possible shapes such as triangular or quadrilateral. The triangular elements are most popular since they are easy to construct, and they can easily conform to structures with irregular boundaries. Figure 5.2 shows the division of an elliptical domain into triangular elements. Notice that because of the straight edges of the elements, as shown in Figure 5.2, the smooth boundary of the two-dimensional domain is approximated as a polygon.

Once the finite elements are described, the unknown function, or field distribution, u, is described over each element using approximation or basis functions. (This is performed in a manner analogous to the basis functions used in the MoM technique.) Over each element, the field is expanded as a combination of linear two-dimensional basis functions. Such basis functions constitute the simplest interpolatory functions used in FEM. However, despite the simplicity of these functions, they have proven to be very powerful in effectively characterizing many functions describing natural phenomena.

A linear two-dimensional function in x and y defined over a triangular element is fully characterized by its values at three vertices of the triangle. The potential u over a single triangular element can be expressed by:

$$u(x,y) = a + bx + cy \tag{5.11}$$

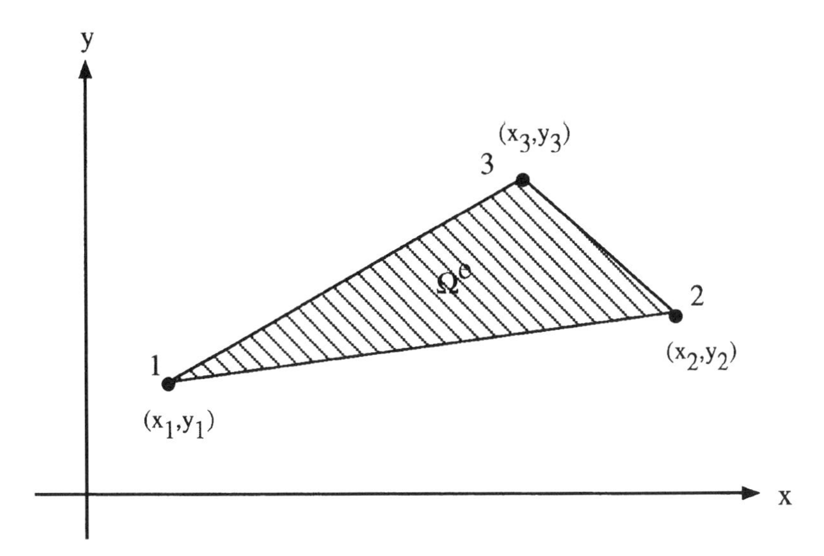

Figure 5.3 Triangular Element in the x,y Plane Having an Area Ω^e

The next step is to describe the potential over the element in terms of the unknown potential values at the three vertices or nodes of the element. Assuming the potential at the three vertices (x_1,y_1), (x_2,y_2) and (x_3,y_3) are given by u_1, u_2 and u_3, respectively (see Figure 5.3), simple algebraic manipulations yield:

$$u = \sum_{i=1}^{3} u_i \, \psi_i \, (x,y) \tag{5.12}$$

where:

$$\psi_1 = \frac{1}{2A} (x_2y_3 - x_3y_2) + (y_2 - y_3)x + (x_3 - x_2)y \tag{5.13}$$

$$\psi_2 = \frac{1}{2A} (x_3y_1 - x_1y_3) + (y_3 - y_1)x + (x_1 - x_3)y \tag{5.14}$$

$$\psi_3 = \frac{1}{2A} (x_1y_2 - x_1y_1) + (y_1 - y_2)x + (x_2 - x_1)y \tag{5.15}$$

The functions ψ_i, where i = 1, 2, 3, are shown in Figure 5.4. A is the area of the triangular element.

When the piecewise planar approximation of the potential is described over the entire domain, the function takes the shape of a jewel-

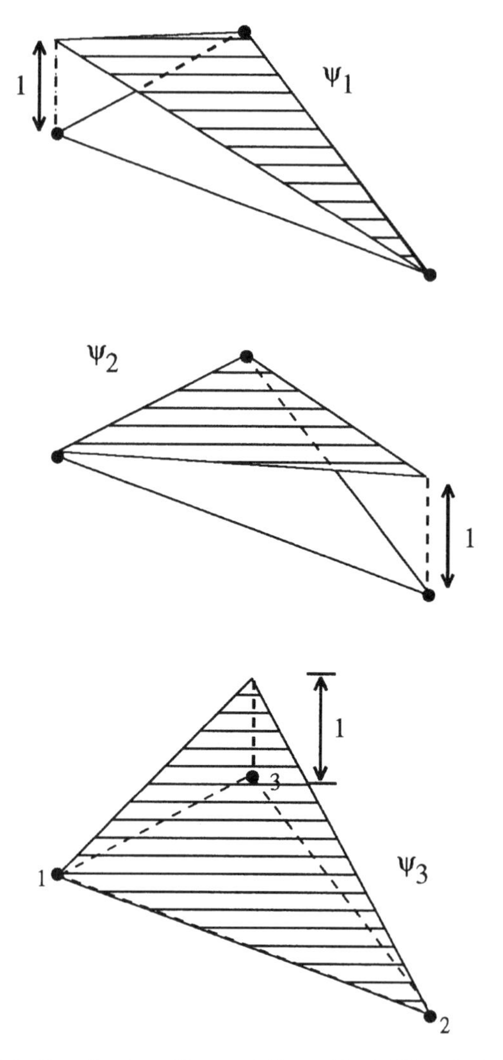

Figure 5.4 Base Functions Described Over a Single Triangular Element

faceted surface. Since each two adjoining elements share the same vertices, and since the approximation of the potential along each edge is a linear function in x and y, the continuity of the potential across element edges is satisfied.

When the potential is described as a piecewise-planar expansion of the field over triangular elements, the elements are referred to as first-order triangular elements. Note that first-order elements involve only three potential or nodal values at the corners of the triangle. Higher-order triangular elements are constructed by expanding the field using piecewise polynomial functions that involve a higher number of nodes, requiring interpolatory basis functions that are polynomials of higher powers of x and y.

5.3.1 Creating the Finite Element Matrix

The unique description of the field or potential in terms of piecewise-planar functions as in (5.12), i.e., a linear summation, allows the functional in (5.10) to be calculated by simply adding the contributions from each element. The functional contribution of a single element is given by:

$$F^{(e)}(u) = \int_{\Omega^e} |\nabla \sum_{i=1}^{3} \psi_i(x,y)|^2 \, d\Omega$$
$$= \sum_{i=1}^{3} \sum_{j=1}^{3} u_i \int_{\Omega^e} |\nabla \psi_i \cdot \nabla \psi_j| \, u_j \, d\Omega \tag{5.16}$$

The presentation can be made more concise by defining:

$$S_{ij}^{(e)} = \int_{\Omega^e} |\nabla \psi_i \cdot \nabla \psi_j| \, d\Omega \tag{5.17}$$

which, after some algebraic manipulations, reduces to

$$S_{12}^{(e)} = \frac{1}{4A} (y_2 - y_3)(y_3 - y_1) + (x_3 - x_2)(x_1 - x_3) \tag{5.18}$$

$$S_{23}^{(e)} = \frac{1}{4A} (y_3 - y_1)(y_1 - y_2) + (x_1 - x_3)(x_2 - x_1) \tag{5.19}$$

$$S_{31}^{(e)} = \frac{1}{4A} (y_1 - y_2)(y_2 - y_3) + (x_2 - x_1)(x_3 - x_2) \tag{5.20}$$

The functional of a single element can now be expressed in the following matrix notation:

$$F^{(e)} = \mathbf{U}^T \, \mathbf{S}^{(e)} \, \mathbf{U} \tag{5.21}$$

Here, $\mathbf{U}$ is a 3×1 column vector matrix containing the nodal potential values given by:

$$\mathbf{U} = \begin{bmatrix} u_1 \\ u_2 \\ u_3 \end{bmatrix} \tag{5.22}$$

and $\mathbf{U}^T$ is the transpose of $\mathbf{U}$ defined as:

$$\mathbf{U}^T = [u_1 \ u_2 \ u_3] \tag{5.23}$$

5.3.2 Matrix Assembly

Assume that the domain or space of the problem is fitted, by a single finite element. Then the only task remaining in the solution procedure is to find the potential nodal values that correspond to the minimum of the functional $F^{(e)}$. The possibility of having only one element to describe the domain is unrealistic. Furthermore, boundary conditions have not been imposed yet, and without boundary nodal values, the potential goes to zero. Therefore by proceeding one step further, the domain can be assumed to contain only two triangular elements as shown in Figure 5.5. The functionals corresponding to each of these two elements are given by:

$$F^{(e1)} = \mathbf{U}^{(e1)^T} \, \mathbf{S}^{(e1)} \, \mathbf{U}^{(e1)} \tag{5.24}$$

$$F^{(e2)} = \mathbf{U}^{(e2)^T} \, \mathbf{S}^{(e2)} \, \mathbf{U}^{(e2)} \tag{5.25}$$

The sum of these two functionals is given by:

$$F^{(e1)} + F^{(e2)} = \mathbf{U}^{(e1+e2)^T} \, \mathbf{S}^{(e1+e2)} \, \mathbf{U}^{(e1+e2)} \tag{5.26}$$

where $\mathbf{U}^{(e1+e2)}$ is a 6×1 matrix given by:

$$\mathbf{U}^{(e1+e2)} = \begin{bmatrix} u_1^{e1} \\ u_2^{e1} \\ u_3^{e1} \\ u_1^{e2} \\ u_2^{e2} \\ u_3^{e2} \end{bmatrix} \tag{5.27}$$

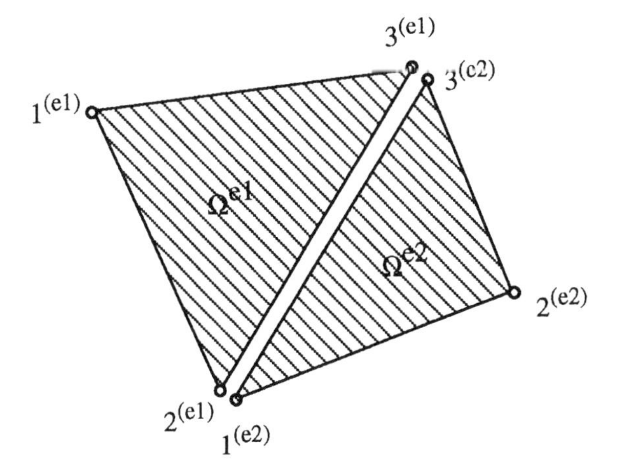

(a) Local numbering

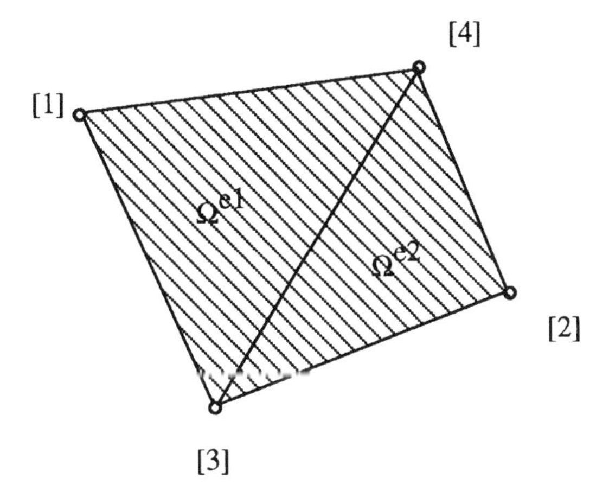

(a) Global numbering

Figure 5.5 Global and Local Numbering of Elements

and

$$\mathbf{S}^{(e1+e2)} = \begin{bmatrix} \mathbf{S}^{(e1)} & 0 \\ 0 & \mathbf{S}^{(e2)} \end{bmatrix} \tag{5.28}$$

When the elements are assembled, the local numbering scheme that references the nodes will no longer be useful and it must give way to a global numbering scheme. Notice that in Figure 5.5, the two elements share the nodes $2^{(e1)}$, $3^{(e1)}$, $1^{(e2)}$, and $3^{(e2)}$. When the elements are assembled, each of the shared nodes, including all other nodes, will have a unique global number. The relationship between the local numbering and global numbering can be presented as a matrix transformation. Let:

$$\mathbf{U}_{local} = \begin{bmatrix} u_1^{e1} \\ u_2^{e1} \\ u_3^{e1} \\ u_1^{e2} \\ u_2^{e2} \\ u_3^{e2} \end{bmatrix} \tag{5.29}$$

and

$$\mathbf{U}_{global} = \begin{bmatrix} u^{[1]} \\ u^{[2]} \\ u^{[3]} \\ u^{[4]} \end{bmatrix} \tag{5.30}$$

then:

$$\mathbf{U}_{local} = \mathbf{C}\mathbf{U}_{global} \tag{5.31}$$

where $\mathbf{C}$ is a connectivity matrix that can be determined by simple inspection of Figure 5.5:

$$\mathbf{C} = \begin{bmatrix} 1 & 0 & 0 & 0 \\ 0 & 0 & 1 & 0 \\ 0 & 0 & 0 & 1 \\ 0 & 0 & 1 & 0 \\ 0 & 1 & 0 & 0 \\ 0 & 0 & 0 & 1 \end{bmatrix} \tag{5.32}$$

This transformation is now substituted in (5.26) and gives a functional representing the connected system of two elements:

$$F = F^{(e1)} + F^{(e2)} = \mathbf{U}_{global}^{T}\, \mathbf{S}\mathbf{U}_{global} \tag{5.33}$$

where $\mathbf{S}$ is a 4×4 matrix given by:

$$\mathbf{S} = \mathbf{C}^{T}\mathbf{S}^{(e1+e2)}\mathbf{C} \tag{5.34}$$

When many elements are used as in a more typical application of FEM, the elements are connected in the same manner as above. Each node of the triangles is connected to only a few neighboring nodes, with the connection being the edges of the surrounding triangles. The number of nodes that are directly connected to the ith node determines the number of nonzero elements in the final assembled matrix. This gives a very sparse matrix system with zeros occupying most of its elements. This resulting matrix can be solved very efficiently using any of a wider variety of special solvers that exploit the sparsity of the system matrix.

5.3.3 Matrix Solution

Equation (5.33) is a functional describing the discrete system obtained using the finite elements and the interpolating or approximating functions. However, once the potential in the domain of the problem is transformed into discrete variables (from a continuous analytic function), the functional F in essence is transformed into a multivariable function having the unknown nodal values as its variables:

$$F = f(u_1, u_2, u_3, \ldots) \tag{5.35}$$

A function has a minimum when its first derivative with respect to all the independent variables is zero. Therefore, the minimum of F is found by satisfying the matrix equation

$$\frac{\partial F}{\partial u_i} = 0; \qquad for\ i = 1, 2, 3 \tag{5.36}$$

5.4 Solving the Two-Dimensional Helmholz Wave Equation

In the above discussion, the FEM theory was presented and applied to solving the Laplace equation. In this section, we discuss the Helmholz wave equation, which appears in many problems in electromagnetic analyses in which the solution domain is the entire open-region of spaces. The procedure of the FEM as outlined above is followed to obtain a matrix equation for the system; however, using FEM to solve open-region problems involves additional considerations, namely that of imposing an outer radiation or absorbing condition. This aspect will be the focus of the following discussion.

5.4.1 Variational Form for the Helmholz Equation

The inhomogeneous Helmholz wave equation in two-dimensional space is given by:

$$(\nabla^2 + k(x,y)^2)u(x,y) = g \tag{5.37}$$

where $k(x,y)$ is the wave number in the medium, and g is a known excitation function.

The Helmholz equation describes the radiation of waves for Transverse Electric (TE)-polarized or Transverse Magnetic (TM)-polarized fields. Here, the use of the FEM is demonstrated in solving the Helmholz equation for the TM polarization where u represents the E_z). The remaining two field components H_x and H_y can be found from E_z, using Maxwell equations.

Two major features distinguish the Helmholz equation from the Laplace equation. The first is the presence of a known excitation function g. In the FEM context, this function is represented using finite elements in a manner identical to the representation for the unknown function:

$$g = \sum_{i=1}^{3} g_i \, \psi_i(x,y) \tag{5.38}$$

The second feature is the presence of the wavenumber $k(x,y)$, which can vary according to the constitutive parameters of the medium. If k is a constant in the entire space, then the medium is homogeneous and

k is simply a number. However, if k varies according to some distribution, k is represented as a piecewise-constant function, which means that k will have a single value over each element. By doing this, the FEM method incorporates the inhomogeneity of the medium without any complication.

The construction of a functional for (5.37) follows (5.3):

$$F(u) = \int_{\Omega} (\nabla^2 u + k^2 u)u - 2\, gud\Omega \tag{5.39}$$

This form can be simplified by invoking Green's theorem which gives the following identity:

$$\int_{\Omega} (u\nabla^2 u + \nabla u \cdot \nabla u)\, d\Omega = \int_{\Gamma} u\frac{\partial u}{\partial n}\, d\Gamma \tag{5.40}$$

Substituting (5.37) in (5.40), we have:

$$F(u) = \int_{\Omega} \nabla u \cdot \nabla u - k^2u^2 - 2gu\, d\Omega - \int_{\Gamma} u\frac{\partial y}{\partial n}\, d\Gamma \tag{5.41}$$

The resulting functional has a strong resemblance in form to the functional constructed earlier for the Laplace equation, however, it has an important difference in the presence of the line integral term on the right-hand side of (5.41).

5.4.2 Absorbing Boundary Conditions

In radiation problems where the radiation takes place in an open region, such as a line source radiating in free space, the domain of the problem extends to cover the entire space. To cover the entire space with finite elements would require infinite computer resources. Therefore, the solution domain of the problem has to be truncated to a finite size by using a domain termination boundary, as shown in Figure 5.6. Over this terminal boundary, a radiation condition, or an Absorbing Boundary Condition (ABC) must be enforced such that no reflection takes place as the waves impinge on the boundary from the inside. The ABC is

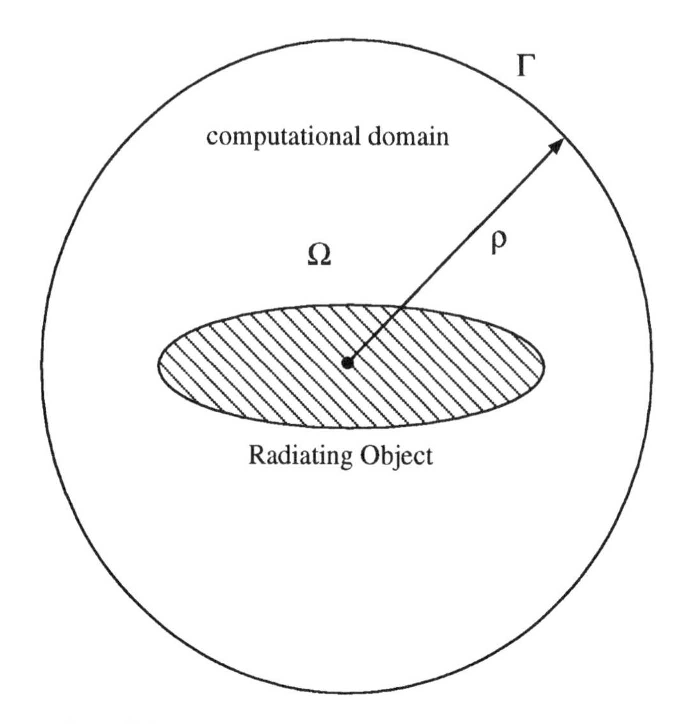

Figure 5.6 FEM Solution of an Open-Region Radiation Problem

enforced on the normal derivative at the terminal boundary, which appears in the right-hand side integral of (5.41).

The simplest ABC is the Sommerfeld radiation condition which is enforced at the outside circular boundary, and is expressed as:

$$\frac{\partial u}{\partial n} = -jku(\rho) \tag{5.42}$$

The Sommerfield condition assumes that the radiating waves are purely traveling, and therefore, for this condition to work effectively, the terminal boundary has to be located at a considerable distance from the radiating object (typically several wavelengths) such that the evanescent waves are not present. Therefore, in general, the Sommerfeld condition has limited value, since it would result in very large mesh size leading to excessive memory requirements.

An ABC that can be brought closer to the object than the Sommerfeld condition is the popular Bayliss-Gunzburger-Turkell (BGT) ABC which is similarly enforced on a circular boundary and is expressed as:

$$\frac{\partial u}{\partial n} = \alpha(\rho)u + \beta(\rho)\frac{\partial^2 u}{\partial \phi^2} \tag{5.43}$$

where:

$$\alpha(\rho) = -jk - \frac{1}{2\rho} - \frac{j}{8k\rho^2} + \frac{1}{8k^2\rho^2} \tag{5.44}$$

$$\beta(\rho) = \frac{j}{2k\rho^2} - \frac{1}{2k^2\rho^3} \tag{5.45}$$

Applying the variable transformation:

$$\frac{\partial u}{\partial \phi} = \frac{\partial u}{\partial l}\rho \tag{5.46}$$

and inserting (5.43) into (5.41), the final functional form becomes:

$$F(u) = \int_\Omega \nabla u \cdot \nabla u - k^2u^2 - 2gu \, d\Omega - \int_\Gamma \alpha(\rho)u^2 + \beta(\rho)\rho^2 u\frac{\partial^2 u}{\partial l^2} \, d\Gamma \tag{5.47}$$

Notice that this ABC involves the second order derivative in the l. This derivative is implemented by using finite difference approximation at each boundary node. To show how this is accomplished, consider Figure 5.7, which shows three nodes on the outer circular boundary terminal. These nodes are denoted as u_1, u_2, and u_3 and are positioned to have a uniform angular separation of $\Delta\phi$. The finite difference approximation gives

$$\frac{\partial^2 u}{\partial l^2} = \frac{u_1 - 2u_2 + u_3}{(\Delta l)^2} \tag{5.48}$$

where Δl is the actual distance (arc) between two adjacent nodes and $\Delta l = \rho\Delta\phi$.

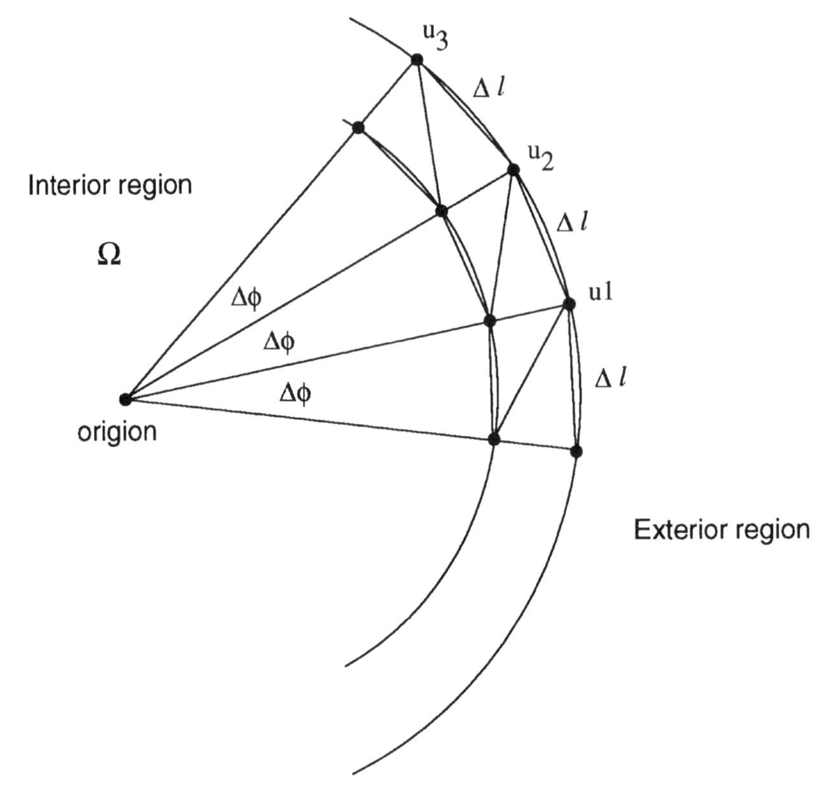

Figure 5.7 Nodes on the Outer Circular Boundary

5.4.3 Field Extension

In most EMI/EMC modeling problems, the field is sought at a distance from the device being tested. For instance, in FCC emissions testing of devices, the field is desired at a distance of 3 or 10 m. Since the FEM solution provides field values that are enclosed within the terminated domain of the problem, the field at a location exterior to this FEM domain, as illustrated in Figure 5.8 has to be calculated indirectly from the FEM solution region. For this, a field extension technique is used that is analogous in theory to the field extension procedure used in FDTD analysis.

The most convenient field extension procedure that can be used in two-dimensional space builds upon the general description of the radiated field in two-dimensional cylindrical geometry. Suppose that the

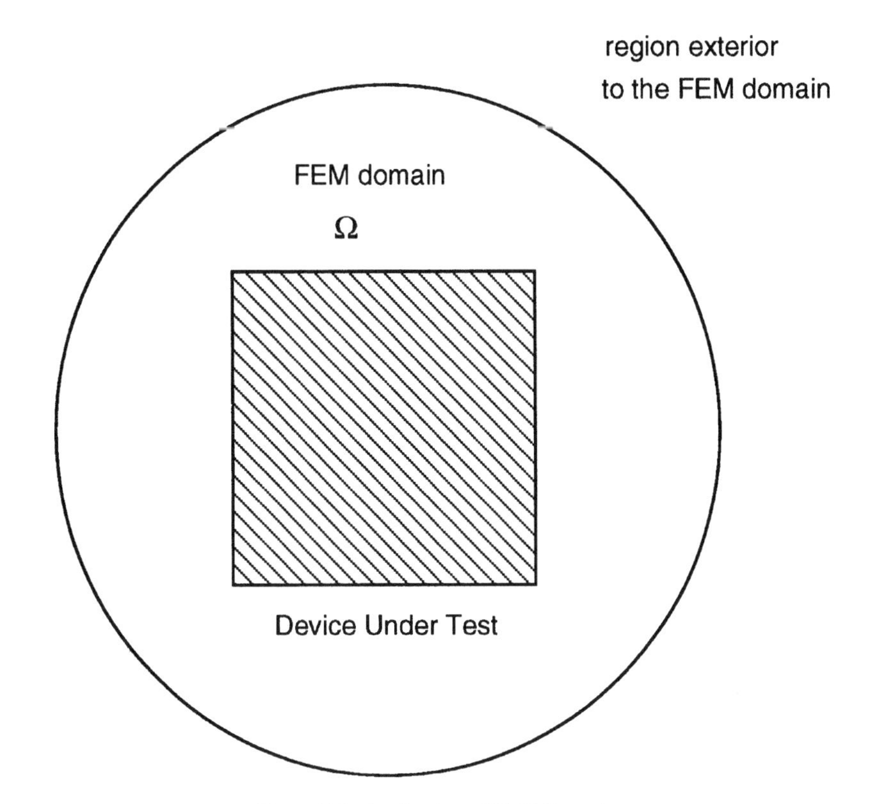

Figure 5.8 Application of Field Extension

structure to be analyzed is enclosed within a circular boundary, as shown in Figure 5.6, then the field at any point that lies on or exterior to the boundary can be represented by it Hankel function Fourier expansion

$$u(\phi,\rho) = \sum_{n=-N}^{N} a_n H_n^{(2)}(k\rho)e^{jn\phi} \tag{5.49}$$

where $H_n^{(2)}$ is Hankel function of the second kind representing outgoing waves [9], provides an excellent background on the application of Hankel functions in electromagnetic theory.

The only unknowns that are needed to fully describe the field are the coefficients a_n. These coefficients are determined as follows: In the finite element mesh, we position the boundary nodes along the outer circular boundary, $\rho = b$ such that they are separated by the distance Δl, as shown in Figure 5.7. After using the FEM to solve for the field at these boundary nodes, we express the solution at the outer circular boundary $\rho = b$ using the representation in (5.49):

$$u(\phi,\rho)_{\rho=b} = \sum_{n=-N}^{N} a_n H_n^{(2)}(k\rho)_{\rho=b} e^{jn\phi} \tag{5.50}$$

In (5.50), the left-hand side is a known function obtained from the FEM solution, whereas the right-hand side is simply a Fourier series expansion, where the Fourier coefficients are essentially the product $a_n H_n^{(2)}(k\rho)_{\rho=b}$. Following Fourier theory, we have

$$a_n\, H_n^{(2)}\,(k\rho)_{\rho=b} = \frac{1}{2\pi} \int_0^{2\pi} u(\phi,\rho)_{\rho=b}\, d\phi \tag{5.51}$$

The integral on the right-hand side of (5.51) is approximated using the rectangle rule which makes the assumption that the function u along the boundary is piecewise-constant. Assuming there are a total of M outer boundary nodes, we have

$$a_n = \frac{\frac{1}{M} \sum_{m=0}^{M} u(m) e^{\frac{-j2\pi nm}{M}}}{H_n^{(2)}(k\rho)_{\rho=b}} \tag{5.52}$$

5.5 Numerical Considerations

One of the most versatile features of the FEM is the flexibility allowed in choosing a mesh that is tailored to suit the physical features of the problem. The mesh density is generally chosen to capture the rapid variation of the field in any part of the solution space. For instance, in the shielded transmission line case shown in Figure 5.1, accurate determination of the field variation in the close proximity of the center conductor entails the construction of a dense mesh around the conductor strip, as shown in Figure 5.9. Where the field is expected to have a slower variation with space, a coarse mesh can be chosen. This flexibility allows for tailoring the memory requirements to be sufficient enough for meeting the objectives of the analysis.

The variable mesh density can also be used to a great advantage in wave propagation or scattering problems. It is important to emphasize

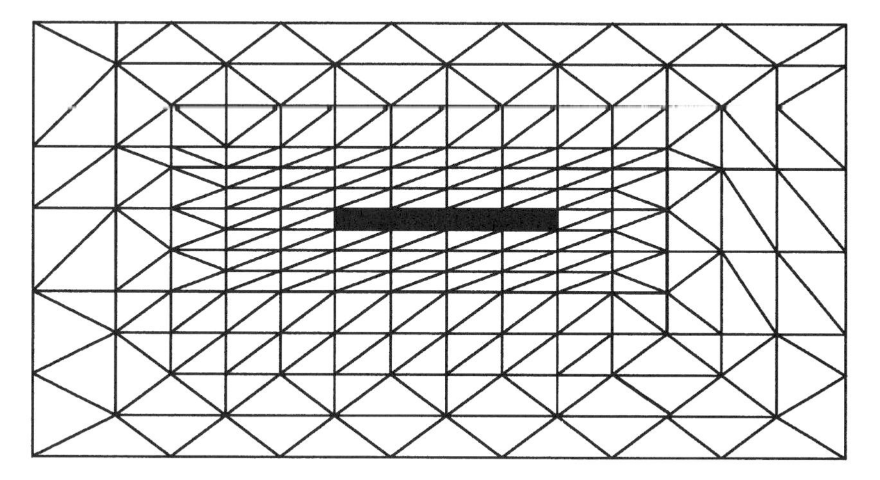

Figure 5.9 FEM Mesh for a Shielded Transmission Line Structure

that in wave propagation problems, the wavelength corresponding to the operating frequency dictates the mesh density. To obtain good accuracy in most problems, it is typical to chose a mesh density that does not allow the number of nodes in any linear dimension to be less than 10 nodes.

The unique construction of finite elements, and especially the way each node interacts with adjacent nodes gives the FEM one of its most primary features. When the nodes are given their global numbers, each row in the final matrix corresponds to a single node in the mesh. Because each node is linked to only few adjacent nodes, the matrix elements in that row are all zero, except a few matrix entries—those that correspond to the connected nodes. When all the elements are assembled, the resulting matrix is very sparse, meaning that most of its entries are zeros. Sparse matrices are quite advantageous; not only can they be solved much faster than dense matrices, but they also can be stored in computer memory very efficiently (a large number of the zeros of the matrix are not stored).

5.6 Summary

This chapter introduces the FEM as it is applied to solve partial differential equations encountered in electromagnetic theory. The discussion focused on the solution of problems in two-dimensional space only.

The solution of either scalar or vector fields in three-dimensional space can be a laborious task not only in the formulation stage of the problem, but also, and more critically, from the practical engineering aspect. This is because the creation of three-dimensional finite element models require considerable skill and training. This is especially the case when solving open-region radiation or scattering problems since mesh truncation techniques tend to perform poorly and would typically demand a large "white" space beyond the structure under study.

References

1. G. Strang and G.J. Fix, *An Analysis of the Finite Element Method,* Prentice-Hall, Englewood Cliffs, NJ, 1973.
2. P.P. Silvester and R.L. Ferrari, *Finite Elements for Electrical Engineers,* Cambridge University Press, New York, 1986.
3. J. Jin, *The Finite Element Method in Electromagnetics,* John Wiley & Sons, New York, 1993.
4. O.C. Zienkiewicz, *The Finite Element Method,* McGraw-Hill, New York, 1977.
5. R. Wait and A.R. Mitchel, *Finite Element Analysis and Applications.* John Wiley & Sons, New York, 1985.
6. B.H. McDonald and A. Wexler, "Finite-element solution of unbounded field problems," *IEEE Trans. Microwave Tech.*, vol. 20, pp. 841–847, Dec. 1972.
7. A. Bayliss, M. Gunzburger, and E. Turkel, "Boundary conditions for the numerical solution of elliptic equations in exterior regions," *SIAM J. Appl. Math.*, vol. 42, pp. 430–451, April 1982.
8. O.M. Ramahi, "Boundary Conditions for the Solution of Open-Region Electromagnetic Radiation Problems", Ph.D. Thesis, University of Illinois at Urbana Champaign, Urbana, IL, 1990.
9. R.F. Harrington, *Time-Harmonic Electromagnetic Fields,* McGraw-Hill, New York, 1961.
10. A. George and J. Liu, *Computer Solutions of Large Sparse Positive Definite Systems.* Prentice-Hall, Englewood Cliffs, NJ, 1981.

Chapter 6

Preparation for Modeling

6.1 The EMI/EMC Problem

Electromagnetic compatibility is a state in which all electromagnetic sources and potential victims can coexist without any unwanted interference. Regulatory limits have been established to reduce emissions to acceptable levels and to eliminate interference. Predicting the performance of a given system relative to these limits is difficult. Modeling can be used to help predict the system's performance.

Traditionally, EMI/EMC designs are based on the previous experience of the engineer responsible for the product's design. To this experience is brought a wide range of supplemental design aids which include both design rules and analytic expressions. The rules are used to minimize the potential for unwanted emissions, usually only providing [only] the preferred approach and not a specific level of noise suppression. More recently, artificial intelligence or "expert systems" have been used to ensure that design rules are consistently implemented.

Computational electromagnetics applied to EMI/EMC modeling is also a recent development and represents an even greater expansion of the designer's resources and capabilities. Through the use of modeling it is possible to both quantify the expected reductions of noise suppression from the use of design rules, and to expand the databases for expert systems. This is an important step, as it permits the designer to concentrate on the areas most likely to cause problems, eliminating wasted time and resources in an effort to minimize areas of lower risk. Further, modeling can be used to determine how much improvement is possible given a perfect implementation of the control technique; the value in this approach is to show when additional or alternative measures will be needed.

Through the use of modeling, databases for expert systems can be made even more inclusive by providing the power of computational electromagnetics in an easily available form. It is clear that for today's high-tech environment, EMI/EMC modeling needs to be integrated into an expanded EMI/EMC design process to ensure that work can be done both accurately and in a timely manner.

6.1.1 The Problem

The first steps in developing a computer model for EMI/EMC design or analysis are to define the problem and to determine what can be obtained through the use of computational electromagnetics modeling tools. When defining the problem, it is always best to go back to the fundamental elements of the situation, and let the model fill in the details. It has become common practice for EMI/EMC engineers to try to double guess the physics of a problem. When using computational models it is more important to provide solid starting data and to know what is needed from those data. It is also necessary to have a good understanding of the how the modeling tools available can be best utilized to address the problem.

All problems can be broken down into three separate parts: the source of the energy, its radiation mechanism, and the means by which it couples from the source to the radiation mechanism. The source of the energy could be any electrical current which changes with respect to time. For example, a fast rise-time clock trace or data bus could be a source of energy. The radiation mechanism is the means by which the RF energy is transmitted to the measurement antenna (typically at a distance of 10 m from the equipment). This could include EMI leakage through air vents and slots, or common-mode EMI noise conducted onto external cables, and then radiated. Coupling between the source and the radiation mechanism can be via internal radiation, conduction along a PCB reference plane, or a combination of both radiation and conduction.

EMI sources may be characterized by either circuit values or field terms. Two circuit values are required to fully specify a source: amplitude and impedance. The amplitude may be a measure of the source current or voltage. The impedance is required to ensure the proper coupling of energy from the source into the final radiation mechanism. If field terms are used, they may be given in terms of electric or magnetic field strength with the impedance relationship between them (near-field or far-field). In general, for radiated emission and conducted problems, the source will usually be a circuit value, while for immunity to external fields, the source will most usually be the field strength of a plane wave.

As an example, one possible EMI source is the high frequency harmonics generated by a clock circuit which has extremely short rise- and fall times. While it seems appropriate for EMI control to minimize the clock's harmonic content, this is often not possible due to the requirements of today's high-performance processors. Another possible source is the noise on the power and ground reference planes that result from the use of insufficient decoupling capacitors near high speed integrated circuits. Again it may not be possible to eliminate the source of the noise, and EMI modeling would be required to determine the implications of this unwanted RF energy.

The radiation mechanisms of this unwanted energy are seldom clearly defined. EMI results when the unwanted energy from a source is present in unintended places and at unacceptable levels. Often more than one possible radiation mechanism must be investigated by the engineer wishing to model a system. When a shielded enclosure is used to shield emissions from a PCB, the openings (such as air vents, seams, and slots) or wires exiting the enclosure (through connectors) allow energy to escape the enclosure. EMI common-mode signals on an external cable are often the cause of excess emissions, since the cable can significantly extend the effective electrical length of the Equipment Under Test (EUT), resulting in a more efficient radiation system. Non-shielded PCBs can also radiate directly.

Energy couples from source to susceptor circuits through two basic mechanisms; conducted and radiated. These coupling paths are often combined in complex ways. In the case of conducted coupling, the source energy is conducted through unintentional paths to the susceptor circuits. This can be due to a common impedance or to a mutual impedance between the two circuits. Common impedance coupling occurs when a portion of the source current flows through an impedance common to both circuits, resulting in a noise voltage in the susceptor circuit. Since only a small common impedance is sufficient for interference to result, it is possible for the source to be quite distant from the susceptor. In mutual impedance coupling, the source and load elements are considered to be closely coupled through electric and magnetic fields such that a change in the source circuit will have a measurable effect on the susceptor circuit and visa versa. This coupling mechanism requires that the source and load be in close proximity.

An example of conducted coupling through a common impedance is the noise found at an interface port of a computer due to the fast current switching of a device located remotely on the motherboard. In this case the common impedance is that of the power and/or ground reference paths shared by the source device and the interface port. Mutual impedance coupling often occurs where printed circuit traces run in close proximity, as in the case of an interface etch located adjacent to a clock etch with high harmonic content.

The second mechanism by which energy may couple is electromagnetic radiation and pickup. Radiated coupling can be considered a special case of mutual impedance coupling where the source and load are not in close proximity and do not have a significant influence on the behavior of the other. For example, a fast rise-time clock trace could be physically far from an I/O connector, but radiation from this trace could cause unwanted radio frequency (RF) currents to flow in the I/O connector. The effects of unwanted radiated energy can be serious both close to the source and many miles away, especially if the unwanted emission occurs on a communication frequency.

Radiated interference is most often a problem to communication devices due to the high sensitivity of these devices. As switching frequencies increase even short conductors become unintentional antennas, such as a large heat sink on a high-performance microprocessor. Although these heat sinks are often electrically small, they are very closely coupled to the high speed, high energy signals within the integrated circuit and therefore can become significant EMI sources. Such a device can radiate energy over a wide range of frequencies and with considerable efficiency, such that cellular telephone or broadcast transmissions are degraded.

When examining many problems in detail, it will often be found that a combination of coupling mechanisms are at work. The paths taken by interfering energy are often very complex and include intermediate steps. This can result in multiple paths with the above mechanisms in both serial and parallel configurations. This makes the task of identifying specific paths very difficult but also highlights the power of creating a fully detailed computational model which, if properly constructed, will include all paths. When necessary, separate models can be constructed to evaluate the contribution from each path, and determine the most significant path. This allows the EMI/EMC engineer to focus EMI/EMC design activities in the most beneficial area.

In addition to emissions problems, the EMI/EMC engineer must also address the topic of a system or device's immunity to interference from external sources. For regulatory purposes this is often specified by the magnitude of an impinging plane wave on the device under test or by a current injected onto an interface cable or directly onto the enclosure as is the case for electrostatic discharge immunity testing. For these cases the coupling mechanisms are the same but the sources are now external and the susceptors internal to the device under test. The large number of processors used in today's automobiles, which control almost every aspect of the vehicles performance, must be immune from both strong external sources, such as would be experienced when driving near a broadcast station or airport, and from local sources such as cellular phones. The ability to accurately predict the reliability of critical systems is extremely important in a world in which almost everything is controlled electronically.

6.1.2 Application of EMI/EMC Modeling

An unlimited range of EMI/EMC problems need to be addressed. Therefore, it is important to select the proper tool for the task at hand. Before commencing the construction of a model, it is prudent to consider how the modeling tools can be best used to address the problem. There are a number of fundamental approaches to using modeling techniques and, in order to obtain the fullest benefits of modeling, these approaches must be used appropriately for each given problem. The intended use for the computer model will determine which technique is most appropriate to provide the solution with acceptable accuracy and in the simplest possible manner.

EMI/EMC models can be categorized into two types of problems: those used for parameter extraction, and those used to solve field problems. The nature of the problem determines which class of modeling tool will be most appropriate. An EMI/EMC modeling tool consists of a user input interface for entering the problem details, a core comprising one or more computational electromagnetics techniques to provide the solution, and a user output interface to display the resultant data in a suitable form. The main differences between these two classes of tool are usually the user interfaces rather than the computational cores. Therefore, while closely related and even over lapping in some areas, the selection of a modeling tool should be heavily influenced by its intended use so that data can be most easily entered and the results made available in a suitable format.

Parameter extraction is the process of calculating circuit element values that relate conductors to one another for inclusion in circuit analysis. The data obtained are usually in the form of a matrix that relates a number of conductors to each other, having elements of resistance, inductance, and capacitance (RLC). Extracting circuit parameters is a common task for signal integrity engineers. For the EMI/EMC engineer it may be a more complex issue as lower value elements can be of concern. In addition to RLC values, the term *radiation resistance* should be added. Radiation resistance is a common term to antenna engineers since it represents the energy lost in a circuit through radiation. Radiation resistance is frequency dependent and provides no information on the directivity of the radiator. The measure of total radiated energy, determined from the current through this resistance, can be a good first pass indicator of EMC performance.

Field problems are those that relate a source to a radiated electric field value. The source may also be a field value or it may be a circuit value. The determination of a system's emissions is a field problem, and is frequently required by EMI/EMC engineers and many managers who would like clear proof of a problem before committing resources to solve it. This is not a trivial problem and, given the very wide range of variables usually present, close correlation for absolute values should not be expected (see Chapter 9). The emissions from an antenna can be easily obtained. For real systems, however, there are usually multiple source locations, each with their own frequency characteristics. The problem geometry is seldom fixed as interconnecting cables are moved from one test to the next and variations in strays and parasitic elements result in minimizing the degree of correlation possible. For this reason, it is a benefit to include intermediate results in a model that can be measured more clearly. For example, since radiation is a function of current on the exterior of the system conductors, the external problem can be solved separately. In this case, a voltage source can be included in the model at the base of interconnecting cables or across large slots as appropriate.

Although absolute electric field strength prediction is difficult, relative field strength prediction is extremely useful. For example, the effect of changing the size of an air vent slot, or the effect of changing the amount of filtering on a I/O connector can be predicted in a straightforward manner.

Susceptibility to external stimuli is another example of a field problem and can be modeled either directly or through reciprocity. Developing a model that will predict the current induced on a conductor within a system, given an applied external field, requires a modeling technique and a model geometry that permits such an excitation. However, if the problem is reversed and a current source is imposed on the susceptible conductor, the field strength at the location of the original interference source can be determined. This will provide a reciprocal transfer function between the source and susceptor, facilitating the determination of susceptibility.

Experimental work is commonly used in EMI/EMC work. Modeling can be used in combination with this experimental work resulting in both being significantly enhanced. A series of specific measurements will provide solid data on which a model can be developed. Similarly, a model can be created to predict specific experimental results. For example, a model could be developed to predict the noise voltage between an interface cable and an enclosure due to an internal source on a PCB. This voltage can be more easily measured than the corresponding radiated emission; therefore, the comparison between measured and modeled results will be more meaningful.

6.2 Overview of Modeling Options

When designing a computer model to analyze an EMI/EMC problem, the engineer has many options from which to choose. One choice is that computer models may be created to be two- or three-dimensional representations of the problem geometry. Both have their place as useful tools, but there are benefits and limitations for each, which need to be considered. Once the dimensional order of the problem geometry is defined, the problem can then be solved for either the quasi-static or full wave solution.

Quasi-static techniques can be used to solve problems where the geometry is small compared to the wavelength of interest. They solve for the electric and/or magnetic fields independently. To obtain a full solution to Maxwell's equations, for a structure of arbitrary size, a full wave technique is required. Within the various full-wave techniques, the solution can be found in either the frequency or time domain. Each technique has its own strengths, and each has its own weaknesses. Which technique is most appropriate for a particular problem depends on the problem. No one technique will solve all problems that an EMI/EMC engineer is likely to wish to model.

6.2.1 Two- and Three-Dimensional Models

Computer models can be developed using either two dimensions or three dimensions. To fully describe an electromagnetic field in Cartesian coordinates, a total of six field terms are required; one electric and one magnetic term for each of the three dimensions. A three-dimensional model is one in which all six of the field terms are included, while a two-dimensional model is defined as having field terms that vary in only two dimensions; the fields in the third dimension are assumed to be invariant. As a result, of the six field terms required for full three-dimensional models, only three need to be considered in two-dimensional models. Either two electric fields and one magnetic field or two magnetic fields and one electric field are used. The selection will depend on the specific problem.

Working with two-dimensional models is particularly easy when the structure being modeled has a uniform cross section as in the case for a transmission line problem. If the problem is more complex, the three-dimensional model becomes more desirable. Two-dimensional models are invariant in the missing dimension and therefore, for a structure that is fundamentally three-dimensional, it is essential to carefully select the cut at which the model is defined. This makes complex modeling very challenging in two-dimensions; a model that does not properly account for the nature of two-dimensional space can easily produce erroneous results which lead to false conclusions.

Two-dimensional modeling has a number of advantages over three-dimensional modeling. Significantly less computer resources are required for two-dimensional models thereby allowing models to be larger or to run faster than the three-dimensional equivalent model. Similarly, two-dimensional models are excellent teaching tools. The presentation of field plots either as a movie or series of individual frames helps greatly to clarify electromagnetic behavior to student, manager, and expert alike. This makes two-dimensional modeling very attractive for a number of applications.

All computational electromagnetics codes have a user interface with varying complexity that permits the problem to be entered, calculation options to be selected, and output data to be displayed in a variety of ways. When first learning how to use a new computer modeling tool, it is best to use two-dimensional models as they are relatively easy and models can be created quickly. Once familiar with the two-dimensional version of a tool, the added complexity of working with three-dimensional models can be done with greater confidence.

Two-dimensional modeling does have some serious limitations. Figure 6.1 shows a two-dimensional and three-dimensional view of a

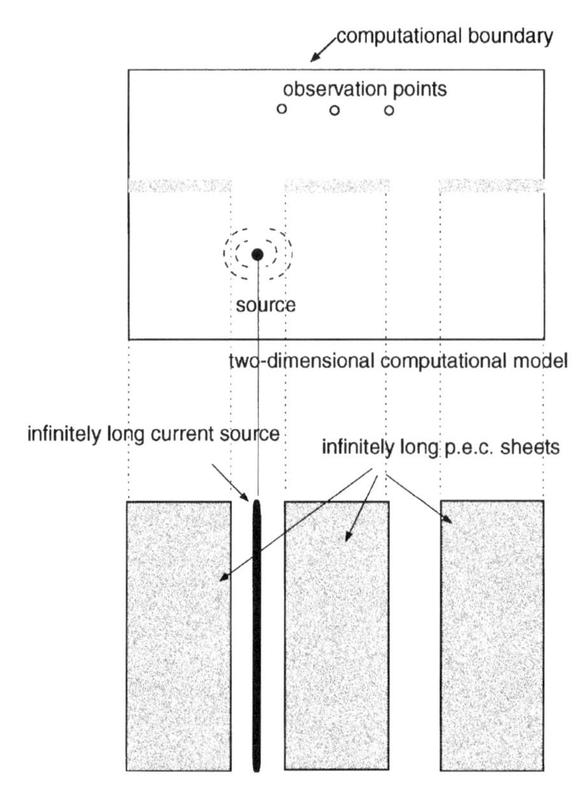

Figure 6.1 Example Comparison of Three-Dimensional Geometry Reduced to Two-Dimensional Representation

problem geometry. In this example, the problem to be modeled is a multilayer circuit board. An internal layer contains a source, and it is necessary to determine how much energy will be received on the top layer of the PCB, where there are a few microstrip conductors on a layer between the source and the outer layer. It is easy to see how the two-dimensional model could be mistaken for a shield problem, rather than a multiple conductor transmission line problem. The two-dimensional model will only be valid for cases in which the components of the model are all long compared to the wavelength of concern in the missing direction, and spacing between the components are very short. It is also necessary to understand how the computational code behaves in the two-dimensional case as energy confined to only two dimensions will not propagate with the same attenuation constants as for those created in three-dimensions. To prevent misunderstandings which are possible with two-dimensional models, it is highly recommended that, until a good familiarity with basic EMI/EMC modeling is obtained, three-dimensional models be used for real problems.

The strengths of three-dimensional modeling lie in its lack of ambiguity through the accurate representation of the problem geometry, and greater flexibility which permits the extension of the computed fields. A three-dimensional model has no spatial restrictions on field behavior and so has a distinct advantage as most problems are three-dimensional in nature. Therefore, no simplifications are required to fully represent the structure of interest.

Creating three-dimensional models is a more complex task and requires careful planning. The degree of detail usually present in a mechanical CAD database is usually much too great for direct use in EMI/EMC models. Therefore, the geometry has to be simplified or recreated for the model. Once the three-dimensional model is created it must be realized that a considerable increase in computational requirements and in the required execution time will be necessary to solve it, as compared to a two-dimensional model that represents a single slice of the problem.

The creation of both two-dimensional and three-dimensional models of complex problems will help the user develop a better understanding of how the two models relate for the case being examined. As only three field components are included for the two-dimensional problem the two-dimensional model must be created so that the appropriate fields are used in the simulation to permit correlation with the three-dimensional model. The accuracy and detail available from the three-dimensional code is well complemented by the speed of the two-dimensional code. Once the correlation is understood, faster and/or larger two-dimensional models can be used to examine the many "what if" cases, and the final selections again can be modeled in three-dimensions to ensure that both accuracy and full details are evaluated properly.

6.2.2 Quasi-Static Techniques

Quasi-static techniques refer to methods that can be used to solve EMI/EMC problems where the geometry is small compared to the wavelength of interest. They assume that there is no interaction between the electric and magnetic fields within the problem space, so they are somewhat less complex than their full wave counterparts. The advantages of this simplification are a reduction in the number of equations to be solved and the resulting minimizing of calculation time.

It is important to understand the limitations when using quasi-static techniques. The primary limitation is that the simulation performed is for low frequencies only; this means that the structure being modeled must be small compared to the shortest wavelength of interest. Quasi-static codes are commonly used to extract circuit parameters for use in circuit simulators. In these cases, either individual or full matrices of inductance, capacitance, and resistance values are required.

6.2.3 Full-Wave Techniques

A full-wave computational technique provides a complete solution to Maxwell's equations within the computational space for all conductors and materials. Full-wave techniques are more complex than quasi-static techniques, but they are also generic in nature and have fewer limitations in their use.

The limitations of practical full-wave techniques will vary from technique to technique and on how detailed a model is constructed. Each full-wave modeling technique has limitations, that is, types of models it can not effectively model. And each full-wave modeling technique has strengths, that is, types of models it can simulate very efficiently and effectively. One major limitation is the frequency range over which the model is valid. This limitation is primarily imposed by how finely the problem is partitioned and how close the elements are to the edges of the computational space. While the valid range may include many decades, errors can be introduced if excitation extends beyond the valid range.

Two options are available in the application of full-wave techniques. The solution can be found in either the time domain or the frequency domain. Time domain techniques use a Fourier transform to provide output data as a function of frequency, when required. Frequency domain codes must be run for each frequency of interest.

6.2.4 Time-Domain Techniques

Time-domain techniques use a band-limited impulse to excite the simulation across a wide frequency range. The result obtained from a time-domain code is the model's response to this impulse. Where frequency-domain information is required, a Fourier transform is applied to the time domain data.

The excitation must be chosen with care to ensure good results. While there are any number of driving wave forms available, the most common is a form of the Gaussian pulse. A simple Gaussian pulse contains energy from direct current (DC) to a defined upper frequency. When modulated onto a sine wave (carrier) the spectrum is symmetrical around the carrier frequency and the upper and lower frequency for the excitation is defined. Another form is the differentiated Gaussian pulse that contains no DC component and has a 6-dB/oct falling value with decreasing frequency and the normal fast roll-off at high frequencies. Specialized pulses consisting of a series of cosines are used where very specific frequency characteristics are needed. For most situations one of the Gaussian pulses is adequate.

Since time-domain techniques excite the model over a wide frequency range, it is very probable that resonances within the model will occur. Whenever resonances occur, there will be extended ringing of the associated currents and fields within the computational domain. In order to obtain the full impulse response, it is required that the fields settle to a steady state. Extended ringing can impose the need for very long simulation times. There are a number of options available which can greatly minimize or even eliminate this problem for practical models. This topic is covered in greater detail in Chapter 7.

6.2.5 Frequency-Domain Techniques

Frequency-domain codes solve for one frequency at a time. This is usually adequate for antenna work and for examining specific issues. Frequency-domain codes are in general faster than their time-domain cousins. Therefore, several frequency-domain simulations can usually be run in the time it would take for a single time-domain simulation. A further benefit to using frequency-domain codes is their capacity to use larger meshes for the lower frequencies, which in turn permits a shorter computation time.

To cover a wide frequency range with frequency-domain codes, a number of simulations are required. It should be noted that there are interpolation techniques available that minimize the number of simulations required. However, these interpolation techniques must be used with care to ensure that a resonance effect is not omitted.

6.3 Selecting A Computational Technique

The application of computational electromagnetics for providing solutions to EMI/EMC problems is still in its early stages; the power of these computational techniques lies in their generic nature. Solutions can be produced for wide ranges of conductor geometries and complex excitations without making any modifications to the core computational code. Problems such as a loaded cavity with a probe feed, array of wires in an antenna, or computer processor installed within a partially shielded enclosure can all be examined using the same codes. The constraints imposed by equations commonly in use need not be a concern. A correctly formed model does not require a set of predetermined conditions for which it is applicable.

Most of today's computational electromagnetics codes provide extremely accurate solutions to the problems they are designed to solve. Choosing the best technique need not therefore be based on its accuracy but rather the ease with which it can be used to address the problems at hand.

Generally, a wide range of computational electromagnetics codes is not accessible to a user, so it is essential to determine what can be done with the tools available. It is also more important for the designer to understand the method by which the computational technique is applied rather than to have a detailed knowledge of the specific technique. Knowing what is wanted and what information is available, it is possible to craft the best model possible with the tools available.

Despite vendor documentation there is no ideal technique or code for all EMI/EMC modeling problems. All techniques have their strengths and weaknesses and typically the more flexible software implementations have fewer safeguards in place to prevent the construction of invalid models. EMI/EMC modeling is seldom a precision task because of the wide range of variables in both the initial problem and the measurement techniques used to verify the real system behavior.

There are three computational electromagnetics techniques in common use for EMI/EMC modeling: the Finite-Difference Time-Domain (FDTD), the Finite Element Method (FEM), and the Method of Moments (MoM). A fourth method, which can be considered a derivative of the FDTD method, is the Transmission Line Method (TLM). TLM will not be covered in any detail as its uses are more specialized and few generic TLM codes exist.

6.3.1 Finite-Difference Time-Domain

The FDTD method[1] is a computational technique used to create volume-based models, i.e., those where the components of the model or the entire model has to be meshed throughout its entire volume. The FDTD code is most commonly implemented in Cartesian coordinates for EMI/EMC purposes. The computational space is divided into small cells and all of the problem components are comprised of these cells. The first thing to consider when creating an FDTD based model is the size of the cells. This sets the granularity of the model, defines the smallest detail that can be included and, as a secondary effect, limits the ultimate size of the problem.

The selection of the cell size to be used for a model is critical. It is determined by either the highest frequency of interest or the smallest detail in the model. A requirement of FDTD in its most usual form is that the maximum cell dimensions are small compared to the shortest wavelength of interest. For EMC work, the cell dimensions should be no larger than 1/10th λ at the highest frequency of interest. There are FDTD codes that permit the use of fewer and larger cells by using higher order differencing, but these are mostly used in specialized applications. The second limit to cell size is the model itself. The cell size must be small enough to resolve the smallest details of concern in the model. If a problem is composed of 1-mm-thick materials and the material thickness is considered important, the cell size must be reduced to a point where there are sufficient cells to resolve the material thickness. Naturally, the cell size must be small enough to resolve all important EMI effects, such as apertures. For a sheet metal enclosure of 1-mm thickness, 1-mm cells might suffice when analyzing apertures of 20 mm or more, but 0.2-mm cells would be required to resolve the fields if aperture sizes of 1 to 5 mm were of concern.

As the total number of cells is limited by the computer memory available and the time it will take to run the simulation, it is good practice to keep cell size as large as possible. The time step that the FDTD code will use is calculated based upon the cell size and material properties. Decreasing cell size and increasing dielectric constant or permeability all decrease the size of the time steps that must be used. Therefore, a model with very small cells will not only require many

---

[1]See Chapter 3 for a detailed discussion of the FDTD technique.

more cells for a given size problem but also will have to be run for a significantly larger number of time steps to cover a given frequency range. Further, at very low frequencies, where the cells become extremely small compared to the wavelength, errors begin to increase due to the minute differences being calculated by the FDTD code.

Once the cell size is set, a review of the entire problem is required to determine the total size of the computational domain needed. It is necessary to build the complete model within the computational domain and to allow a region of white space between the edge of the model and the boundary of the computational domain. The size of the white space region depends on the specific Absorbing Boundary Condition (ABC) used by the software developer. Different ABCs require more or less white space, and will add various amounts of computational overhead and various amounts of error. Most ABCs require sufficient white space to allow the radiating fields from the model to approach a far-field relationship where the electric and magnetic fields are related by the impedance of free-space. For EMI/EMC purposes, this typically occurs at a distance of about 1/6th λ at the lowest frequency of interest in the model. Once the total number of cells is known the amount of computer memory needed can be calculated based upon the total number of cells times the number of bytes required per cell.

The total number of time steps must be specified for most FDTD software packages. The simulation must be run sufficiently long to ensure all interactions have been completed within the model, but not so long that computational resources are wasted. If a model has little or no resonance effects, the number of time steps can be roughly calculated by the number of cells between the source and the receiving field monitor points. The size of the excitation (in time steps) must be included as well. If resonance effects are present in the model, then the time required for the resonance to completely dampen depends on the Q-factor of the system. A practical approach is to run the model for a short period of time, observe how rapidly the response is being dampened, predict the correct amount of time steps required based on the dampening rate, and rerun the simulation.

Within the FDTD technique it is possible to make each cell have its own material properties. This is very straightforward if the material property is invariant across the frequency range of interest. However, if the material property varies with frequency, as in the case of ferrite materials, the fields in the ferrite cells must be calculated using special convolution techniques, or average values may be used for specific frequency ranges and later combined.

There has been extensive research on how to approximate non rectangular objects within the rectangular FDTD coordinate system. One of the best solutions allows angled or curved components to be approximated by the use of a staircase of cells. Typically 1/10th to 1/20th λ cells are considered sufficient accuracy for EMI/EMC applications. For cases in which resonant structures must be precisely modeled this can be a significant challenge and may require finer cells than would otherwise be needed. EMI/EMC issues rarely require this high level of accuracy, as the available starting data and other variables of the design have much greater variations. When high precision is required, special implementations are available which enforce field behavior diagonally across a cell or series of cells to increase the flexibility of the FDTD technique.

Owing to the implementation of the Yee cell in FDTD, the electric and magnetic fields are computed on an offset grid, one half-cell apart. As a result of this offset it is not possible to exactly define where the edge of a cell is located. If the electric field is on the surface of a conductor then the closest magnetic field components are one half-cell inside and one half-cell outside the surface. This can be a source of error if very few cells are used to create a model where high accuracy is needed.

The FDTD technique has many advantages for solving EMI/EMC problems. It is a "brute force" method in that it solves Maxwell's equations for the computational domain and frequency range specified and requires little in the way of simplification or approximation of the physical problem. The FDTD technique permits very detailed models to be created, with materials whose properties can be specified to the cell level; further, adding detail within the computational domain does not require greater computer resources or run time. Since the FDTD technique is a volume based method, it allows the RF currents within the model to be different on either side of a metal shield. This means that the FDTD technique is particularly useful for simulations involving shielding applications. Since the FDTD technique is a time-domain solution, a wide range of frequencies are solved simultaneously, and any resonances present will not be missed by too-large frequency stepping.

The price paid for the flexibility of the FDTD technique is the need for greater computer resources in terms of memory and processing time.

Time-domain excitation increases the probability of exciting resonance effects that usually extends the required run time. Since the model must be contained within the gridded computational domain volume, long wires and other long thin structures are difficult to model in the FDTD technique.

6.3.2 Finite Element Method

The Finite Element Method (FEM)[2] is also a volume-based technique and is primarily used in the frequency domain. When creating FEM models, the entire computational space is meshed after all the components are created. This is usually done automatically or semi-automatically, with mesh refinements added by the user, if necessary, near points at which greater resolution is required. It is common for codes to iterate through this process until an acceptable level of accuracy is obtained.

The effect of small details is not overly critical in the FEM as the mesh size is not fixed but varies over the computational domain to ensure that adequate resolution is used only where needed. This is very convenient where a few small details exist, as they may be included without concern. However, too many small details will result in a very large problem to solve. If the problem becomes too large for the available computer resources it is best to closely examine which of the fine details are really needed.

As with the FDTD technique, it is possible to use any material properties needed for each of the FEM model components. Further, since FEM modeling is typically done in the frequency domain, it is straightforward to enter the appropriate property value for each frequency.

Meshing for FEM is usually done with triangular or tetrahedral elements. This means that any object is meshed in a polygonal manner, so the staircase and half-cell effects are not issues with FEM models. As an angle of the mesh grows more acute there is a corresponding decrease in accuracy. Therefore, the objects that cause most concern in an FEM model are those with one narrow and one wide dimension, unless a large number of mesh elements are used to represent the object. Mesh granularity is set both by the size of the model components and

---

[2]See Chapter 5 for a more complete description of the Finite Element Method.

by the frequency of interest. Therefore, it is expected that different meshes will result, depending on the frequency of interest.

FEM models benefit from using sparse matrices that allow the solution to be obtained relatively easily, minimizing the required run time. In addition, the use of adaptive meshing techniques optimizes the use of computer resources. As FEM is a frequency domain code, frequency dependent materials can be specified for the specific frequency of concern. One of the main advantages of the FEM is the convenient ability to vary the mesh size as required. This does not eliminate the need to ensure the largest mesh size is small (1/10th λ) at the highest frequency of interest.

The price paid with any frequency domain code is the need for a number of simulations to adequately cover a given frequency range and, unless sophisticated techniques are used to fill in the missing spectrum, it is possible to miss important resonances within a structure. One of the major weaknesses of the FEM is the need to have excessive white space around the model. The resulting high number of unknowns greatly limits the usefulness of the FEM for EMI/EMC applications. This can be ignored for completely enclosed applications, such as waveguide applications. The FEM has the same limitation for long wires as the FDTD technique described earlier.

6.3.3 Method of Moments

The Method of Moments[3] (MoM) is a surface-based technique, i.e. only the component surfaces are meshed. To represent a radiating system using an MoM technique, all the conductors in the problem have to be incorporated. Surface modeling is done by breaking the surface of interest into patches that are small (typically 1/10th λ), compared to the wavelength of concern. While surface modeling is available it often less convenient for EMI/EMC problems than wire frame models. Creating a wire frame representation of the surface is quite convenient and allows long wires to be attached easily. When creating wire frame models it is essential to remember that these surfaces will radiate in all directions, and not just to the "outside."

Small MoM models can be created very easily on a fairly basic portable personal computer. Creating approximate models to help bound

[3]See Chapter 4 for a detailed description of the Method of Moments technique.

a problem is useful—to validate a rule of thumb, or to determine a range of uncertainty for a series of measurements.

The MoM technique is particularly efficient in its use of computer resources since only the surfaces of model elements are meshed. This also results in relatively short simulation times for the single frequency case. Since one part of the solution technique is to compute the currents on all surfaces, it is equally easy to provide both far-field and near-field observation points. The MoM technique is particularly useful when modeling long thin wires, since the volume around the wires is not part of the basic model.

MoM, like FEM, is a frequency domain code and so simulations must be performed at all frequencies of interest. In addition, care must be taken with MoM models as surfaces are single sided which effectively limits the use of MoM models for shielding and aperture problems.

6.4 Elements of an EMI/EMC Model

The goal of modeling is to create representations of real-life problems that can be examined and analyzed by computer resources as an alternative to building a system, exciting it, and measuring the generated fields. Once the problem has been defined, the important physical characteristics must be identified.

All EMI/EMC models can be broken into three parts, the source of radio frequency (RF) energy, the geometry of the model components, and the remaining problem space. Before addressing specific details of the model, it is helpful to get an overview of how the model will be set up.

6.4.1 Sources

The reason there is concern over EMI/EMC problems is that there are intended and unintended RF sources that couple to and drive conductors such that energy is radiated or conducted into areas where it can cause problems to the correct operation of the victim device. Simply put, an EMI source is RF energy that is not fully delivered from a source to its intended load. This is usually in the form of harmonics or sidebands to the required signals, but they may be the intended signal current traveling an unintended path. This can be caused by reflection, common impedances, and a number of other stray or parasitic effects.

Two areas must be addressed when developing an EMI/EMC source model: (1) the physical representation, which must accurately reflect the geometry of concern; and (2) the driving function used to excite the physical source model. For many problems, the source of the RF energy that is expected to radiate is well known. Therefore, defining the source is only a matter of collecting the necessary data. Source specifics may be obtained either from measurements or circuit simulations, and geometry details may be obtained from either physical measurements or assembly drawings.

There are cases in which nonrealistic source models can be beneficial. One such source is a current in space with no conductor to support it. This source is known as an impressed current source and is not affected by the presence of other conductors; it also has no resonances of its own, as in the case of real conductors. This source is particularly useful as it permits the examination of a structure, such as an enclosure's resonances, without specific source effects.

Other options that exist include creating purely evanescent waves, purely traveling waves, or even isolated modes in a given structure. For these special cases, the power of computational techniques is unsurpassed, allowing detailed examination of a structure that would not be possible experimentally.

6.4.2 Physical Source Modeling

Sources may be characterized by their electrical size, distance from materials with which they interact, geometry, and the excitation applied to them. It is apparent that a wide range of sources exist and, with the flexibility available through computational electromagnetics, all can be accommodated. Previously, simplifications were necessary. At one extreme are infinitesimal electrical and magnetic dipoles; at the other, plane waves. Both extremes represent the sources most commonly used for closed-form electromagnetic equations, and both assume that there is no interaction between the RF source and the radiator/susceptor. However, in many situations, the simplification of using these sources cannot be justified as greater detail is needed. For these cases, computational techniques can be applied to create any specific detailed source structure and solve the resultant field equations.

While the sources of all EMI/EMC problems are RF currents, there are many ways of energizing a simulation. The correct selection of the source type is essential in obtaining the correct results. A properly constructed source will contain elements that accurately duplicate the geometry, and also contain a driving function to energize the source appropriately.

Precise location of the source within the model can be very important. In the absence of specific data it is somewhat natural to locate sources centrally on a structure; however, doing so limits the number of modes that can exist. Therefore, unless it is a true representation of the source, the feed point should be offset from the center or other obvious points of symmetry.

There are still many uses for the simple sources. Infinitesimal dipoles, both electric and magnetic, provide a means of energizing a model with the minimum of interactions. This permits the close examination of the coupling mechanism independent of the source. Plane wave sources are specified by immunity standards and so are ideal sources for exciting susceptibility models. All these sources are easily handled with today's computational techniques and should be used when appropriate.

When the source geometry is small compared to the wavelength of interest and distance from other conductors, it is possible to use a loop or dipole as an adequate representation of the source geometry. In these cases small variations such as mounting inserts can be ignored. Where larger conductors are closely coupled to the electrically small source conductor, they must be included fully as such closely coupled conductors are effectively part of the source geometry.

Plane wave sources are especially applicable to susceptibility models. In these models it is necessary to determine the response a device or system has to an applied electromagnetic field. In addition, plane waves are the traditional source for scattering problems. Although not usually of interest to EMI/EMC engineers, it should be realized that scattering analysis is a well developed and closely related field and should be considered a rich resource for further work in the area of susceptibility.

For cases in which the source conductors are electrically large, the problem can be reduced to analyzing an antenna. In situations where electrically large source conductors are located in proximity to other conductors, these latter conductors become parasitic elements to the antenna structure and so must be included in the model. This is of special importance if the correct radiation patterns are to be obtained. For EMI/EMC purposes it is usually not necessary to predict antenna patterns; only the peak fields that they produce at a given distance.

6.4.3 Source Excitation

Two components are required to fully specify circuit model sources: amplitude and impedance. The amplitude can be a voltage or current. The driving function impedance is very important and, if ignored, unpredictable behavior can result. As an example, consider an ideal current source driving an antenna. A constant current flows into it no matter what impedance is presented. As a result, at frequencies where the antenna is anti-resonant (very high impedance), the effect at the source will be a corresponding increase in drive voltage, as necessary to drive the required current.

Figure 6.2 shows the total radiated power from a dipole antenna when fed with three different sources. For ease of examination of the results, all have been normalized to the same radiated power at the frequency of the first resonance. The three sources used for this example are a perfect voltage source, a perfect current source, and a 10-ohm resistive source. While the absolute amplitude values are not correct, it is clear to see that the three sources result in very different frequency

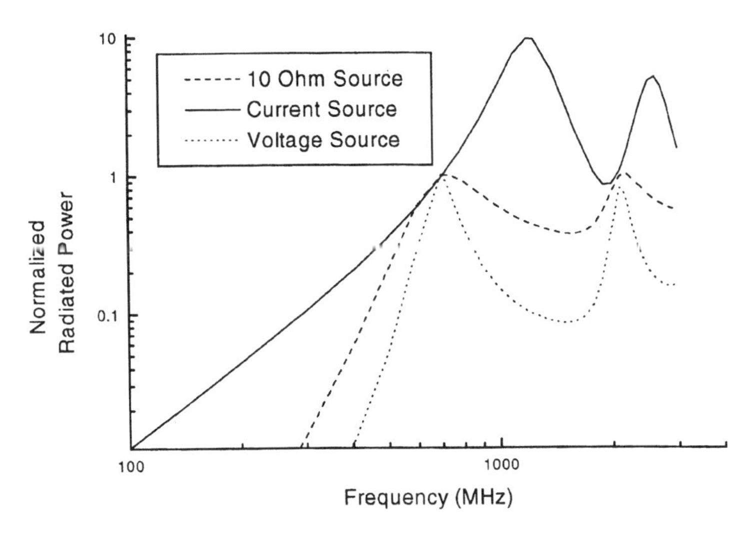

Figure 6.2 Effect on Radiated Electric Field Due to Various Source Impedances for a Dipole Antenna

responses. This demonstrates the importance of obtaining all the source information and not assuming a particular value, such as 50 ohm, as is common for filter measurements and specifications.

Field sources also need to be defined by their amplitude and impedance. While amplitude terms are easy to understand, it is difficult to determine the appropriate wave impedance to be used. For many purposes in EMC work it is easy to use a circuit element source and to let the computational code solve for the field values. Naturally, when a far-field plane wave is the source, the wave impedance of free space (377 ohms) is used.

Computer model excitation can take place in either the frequency domain or the time domain. Frequency-domain excitation is often most easily understood as it is defined by terms compatible with circuit models and consists of a driving voltage or current with an associated source impedance. When working in the frequency domain, each frequency of interest needs to be analyzed. This will require running a number of separate simulations.

Time-domain excitation tends to be a little more complex than frequency-domain excitation and should be selected with care. In both cases, it is recommended that general solutions be sought. As an example, a normalized excitation of 1 volt can be easily scaled to any particular spectral content, permitting a clear understanding of the numerically modeled portion of the problem. Any resonances in the structure can be easily seen whereas, if a frequency-domain source is used, this could be easily missed.

The time-domain excitation waveforms usually are based on the Gaussian pulse. Three excitation pulse examples are shown in the time and frequency domain in Figures 6.3 through 6.5. The first case is a Gaussian pulse; the time domain is shown in Figure 6.3 and the frequency-domain representation is shown in Figure 6.6. This pulse contains a spectrum of energy that is fairly flat up to a given frequency and then drops off rapidly. This pulse has a DC component, which can be a problem in some models. Depending on the geometry of the model, the DC component may result in a net charge left within the computational domain which can in turn lead to instability of the solution.

When a Gaussian pulse is differentiated, the DC component of the pulse is removed. The resulting time domain waveform is shown in Figure 6.4 and its frequency spectrum shown in Figure 6.6. The differen-

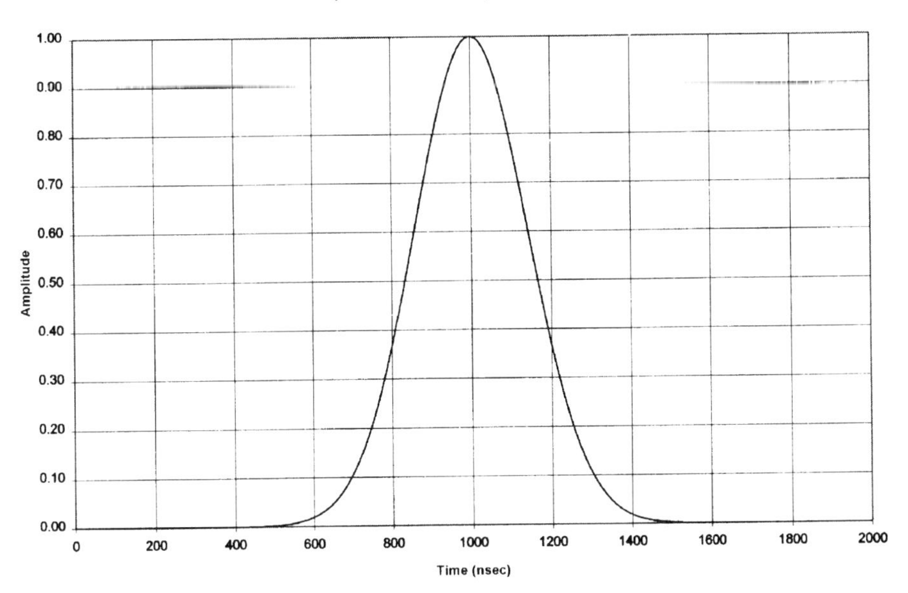

Figure 6.3 Example of a Gaussian Pulse in the Time Domain

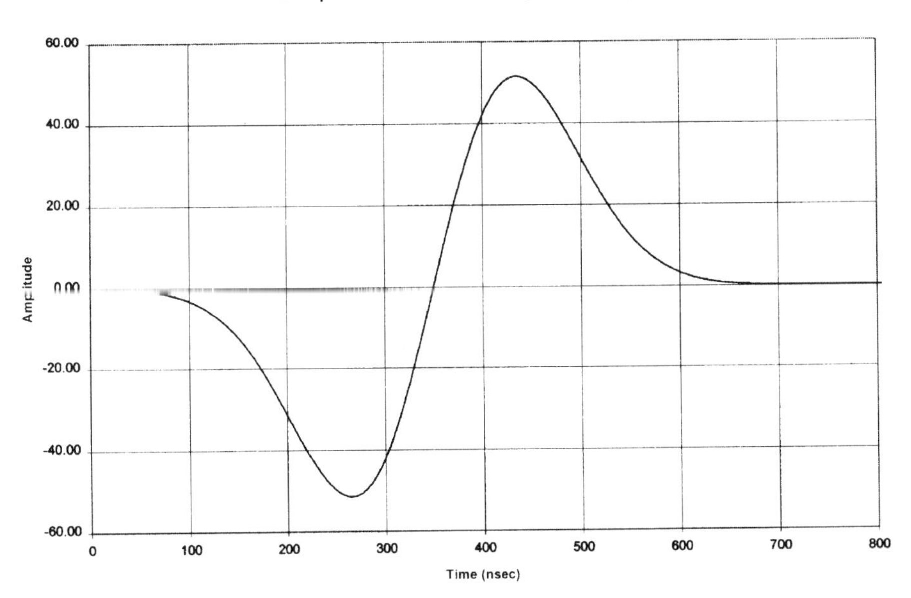

Figure 6.4 Example of a Differentiated Gaussian Pulse in the Time Domain

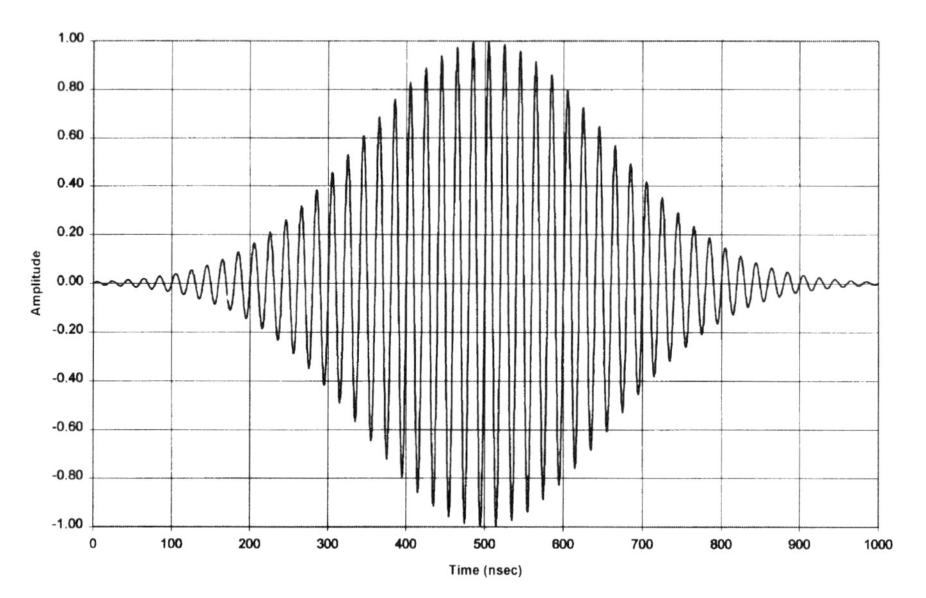

Figure 6.5 Example of a Modulated Gaussian Pulse in the Time Domain

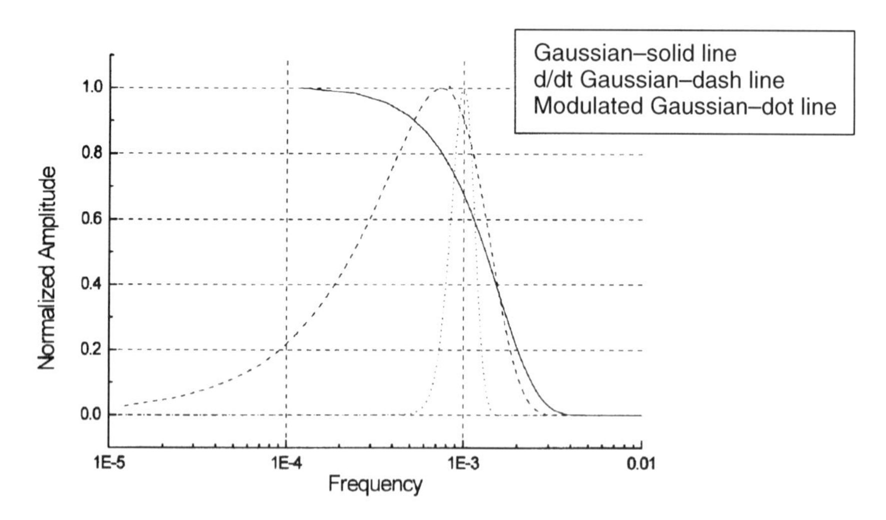

Figure 6.6 Frequency-Domain Representation of Various Gaussian Pulses

tiated pulse has a frequency response that rises at 6 dB per octave to its peak value.

Another very useful excitation waveform is the modulated Gaussian pulse. In this waveform, the Gaussian pulse is modulated with a carrier frequency, and centered in the range of interest. Figure 6.5 shows the modulated Gaussian pulse in the time domain and Figure 6.6 shows its frequency spectrum. The roll-off toward low frequencies is faster than for the differentiated Gaussian pulse, making it a good choice for band-limited applications. In addition to containing no DC component, this pulse has the advantage of minimizing both very low and very high frequency components, both of which may be a source of error under various conditions.

To select the appropriate excitation for a particular simulation, it is necessary to understand how they differ and to have an idea of how the model should respond. For an FDTD model, it is important that no significant excitation be present at frequencies above that determined by the Courant stability conditions for the geometry of interest. The simple Gaussian pulse appears ideal at first glance, as it typically has a flat frequency response over the entire range of interest. However, problems can result from the very low frequency and DC components, which sometimes leave a static charge in the final results. The static charge is easily avoided by using a differentiated pulse. The differentiated pulse provides energy over a wide range of frequencies. When the frequency range of interest is relatively narrow, all the source energy can be focused in the desired frequency range by using the modulated Gaussian pulse, which can be selected to eliminate all unwanted high- and low-frequency components. Custom pulses are only required where the frequency range of interest is very close to the high- or low-frequency stability limits of the model.

6.4.4 Model Geometry

The heart of every EMI/EMC model is the geometry of the problem to be solved. It is rare that a set of mechanical CAD files can be used without a major editing effort. The level of detail contained in such databases results in large numbers of small details such as bend radii, and off set bends. These details would produce a model that is extremely discretized and results in an unacceptably large problem. A less complex representation must be created that includes all the important detail, while avoiding those details that are unnecessary. In addition to the fixed portions of the geometry it is often necessary to include variables such as the range of positions in which a nearby cable, or any other conductor, could be placed.

Together with the geometry of a problem, the properties of all materials used must also be included in the model. For EMI/EMC work most metals can be usually considered as Perfect Electrical Conductors (PECs) and only dielectric or magnetic materials need to be especially defined. For cases in which a PEC cannot be used, the relevant parameters must be added. Depending on the problem being solved, the addition of non PECs may add to the computer resources required.

6.4.5 Completing the Problem Space

For the FDTD and FEM techniques, it is necessary to prevent reflections from the edge of the computational domain, which can affect the accuracy of the simulation. This can be done by either placing a lossy material at the boundary, or simulating a free space environment.

By simulating free space, the computational domain behaves as if it were infinite in extent. The means by which this is achieved are known as mesh truncation techniques or absorbing boundary conditions. These techniques require that extra free space (known as white space) be added around the model components. The white space is used in two ways. Many absorbing boundary conditions require that fields are not changing too rapidly with distance, hence the first use for white space is to provide a buffer from the problem components, where the field can be changing very rapidly, to a region in which the field behavior has settled to some extent. The second use for white space is that all boundary techniques use a small number of field points just inside the computational domain from which the ABC predicts the field behavior on the boundary. The white space is typically created to ensure the ABC is in the far-field from any possible source. For EMI purposes, this is usually set to be about 1/6th λ at the lowest frequency of interest. Figure 6.7 shows an example of an FDTD model with the white space around it.

One aspect of absorbing boundary conditions that is of particular note is the selection of the points used to determine the boundary field values. Some techniques such as Liao use only points orthogonal to the boundary while others such as Mur use both orthogonal and tangential

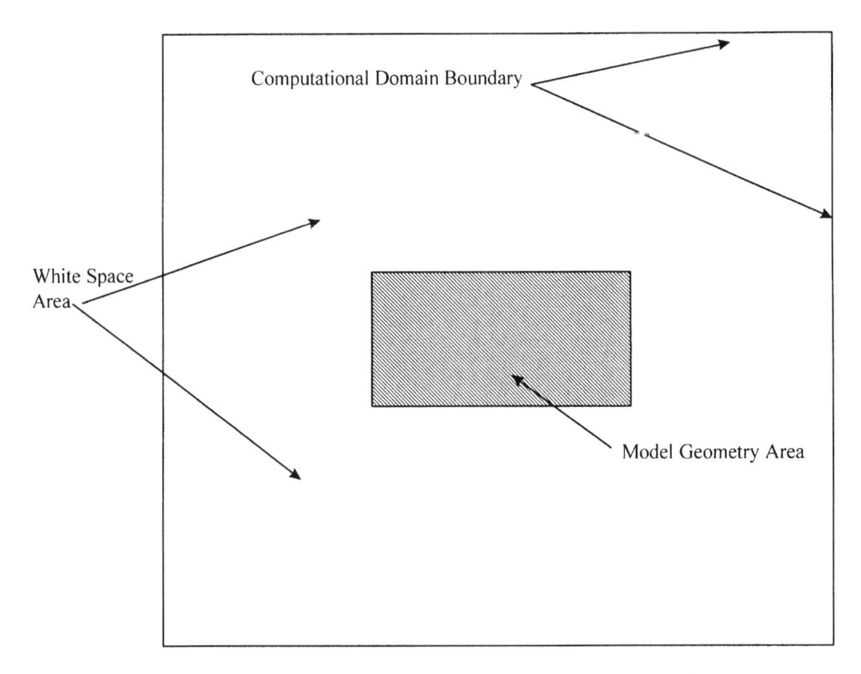

Figure 6.7 Example of White Space Around an FDTD Model

values. This selection can have a significant influence on the range of modeling possibilities. When using an absorbing boundary condition that requires only orthogonal field values, it is possible to permit any PEC to contact the boundary, and the effect is to simulate the PEC extending to infinity. This is of great benefit when shielding issues are being studied as it permits the shield to extend completely across the computational domain and avoids the use of a fully enclosed box.

6.5 Model Goals

The first steps in developing a computer model for EMI/EMC design or analysis are to clearly define the goal of the modeling and to determine what information is available on the system to be modeled. Goals could range from evaluating a system's radiated field strength or susceptibility, to determining various system parameters. Knowing what information is available is also important in defining goals.

Modeling goals should be examined to ensure that they are both definitive and practical. An evaluation of the emissions from system X may be the final goal, but in practical terms the problem will need to be broken down into appropriate parts. Each independent source may require a separate simulation in order that the output data can be understood. Energy can radiate from the system through a variety of unrelated means. Critically thinking through the problem and setting up pieces that can be properly modeled by the computer and understood by the designer is essential. With this step completed and the parts of the system to be modeled defined, the modeling process can proceed most efficiently.

Having considered the issue of what to model, the next step is to determine what information is available on which the models can be based. Computer models are equally capable of solving poorly conceived models as good ones; therefore, accurate starting data are essential. Some of the required data are in a hard form, while other data are less tangible. Although the geometry of a system enclosure, cable placement, or power system might be well defined and available on a mechanical CAD database, the characteristics of the RF energy sources may be lacking in detail during the early phases of the work. The impact of missing data needs to be considered. In many instances an unknown RF source can be replaced by a standardized source, permitting evaluation of coupling and emission levels. This may then be scaled to real source values when they become known.

Once the modeling goals and starting data are defined, there is one more question to address; whether computational techniques are the best means for solving the problem or whether traditional methods are more appropriate. Traditional EMI/EMC estimation techniques are often valid, and useful. Computational modeling tools should be considered as an additional tool in the engineer's tool box, not the only tool. Traditional estimation techniques can also be useful to help validate the results of the numerical model.

6.5.1 Defining Goals

When deciding which data are needed from a simulation, consider all the options available. The output data can be classified broadly into two groups: field parameters and circuit parameters. Electric and magnetic fields can be determined from simulations, as well as voltage, current, and RLC values extracted from the simulation results.

While it is necessary to model specific critical problem areas, it is also important to consider less critical parts of the system. A common cause of EMI/EMC over design is the uncertainty of which areas of the system are critical to proper EMI/EMC design and which may be ignored. As a result, the tightest level of control is typically applied to the total system, rather than to just a few critical portions. Modeling can be used to help identify which parts of the system are critical, and where money and effort can be saved.

When planning the preparation of a model it is necessary to collect all the available information together and determine how comprehensive it is. If important information is not available it may reduce the accuracy or range of models that can be properly created. Computers will solve the problem that is set, even if it was not the problem intended. Often previous experience of similar systems can be helpful in supplying additional information into the model. In cases in which only small changes from similar systems has been made, computer models of the new configurations can be made and combined with the known results from the original system. The differences between the results will be useful output as, for example, when the driving EMI source is changed in the case of a CPU upgrade or a change in limit for a susceptibility test procedure.

6.5.2 Desired Results

The purpose of the EMI/EMC model is to simulate the EMI/EMC problem and predict its behavior. To do this, a means of monitoring the parameters of interest must be implemented. Observation points are the means for achieving this and may be as simple as saving the field strength at a particular location or much more complex, requiring additional processing.

Field observation points may be within or external to the computational space in which the fields of interest are evaluated. For the FDTD and FEM techniques, it is possible to use additional numerical techniques, known as field extension techniques, to evaluate fields at any location external to the computational space. Such field extensions usually use an integration surface around the problem geometry on which sufficient fields are evaluated (see Figure 6.8) such that the overall contribution to any given point outside this shell can be calculated. This adds overhead to the problem and therefore execution time is increased to some extent depending on the number of points required.

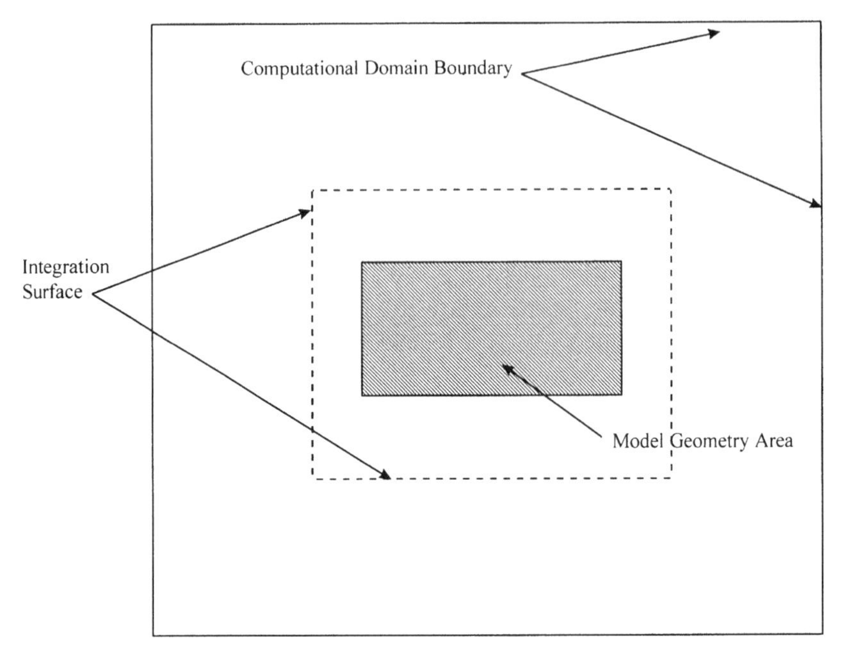

Figure 6.8 Example of Integration Surface Around a Model for Field Extension

Field extension techniques can be used for any point outside the shell containing the model. It has been found that using these techniques can increase accuracy within the computational domain. The increase in accuracy is especially noticed when observation points are located within the computational space and close to its edge. When using a simple field observation point in this location, there are small but real errors due to the reflection from the edge of the computational domain. When using the extended field observation point in the same location, these errors are smaller, as they are derived from the shell which is located further from the edge of the computational domain, and are based on data from the whole shell, rather than a single point.

The MoM technique has no computational domain limitations so there are no truncation issues. Further, as the MoM technique solves for the current distribution over the exterior surface of all conductors, the fields at any location may be calculated with equal ease; therefore, observation points can be freely located.

Specific software implementations of the various modeling techniques may provide preset observation point selections. It is very important to ensure that these are appropriate for use with the selected problem. For example, commercial radiated emission limits are required to be measured over the surface of a cylinder with a given radius, usually 3 m or 10 m, and over a height range of 1 to 4 m. While some codes have these locations as a preselected choice, they may also offer a "maximum field strength" over a hemisphere or sphere that can be much more severe and will also minimize the possibility of matching simulation results to measurements. This effect is further illustrated in Section 9.3.

Circuit element observation points are better referred to as probes in keeping with the circuit model parallel. These probes can be simple, such as voltage, current, and impedance, or more complex such as a matrix of inductance, capacitance, and resistance. Voltage probes require that the user specify two points between which the electric field is evaluated and integrated to give the associated voltage. Current probes can select one or more conductors in which the total current is required. In this case, the magnetic fields are integrated to provide the required current.

Signal integrity codes will often provide the data in other forms, such as S parameters, where such matrices are more commonly used. For EMI/EMC work, these are not typically used.

6.5.3 Problem Geometry

Obtaining details of the problem geometry is a necessary step in producing a model. Within the computational model, the problem geometry will be broken into pieces that are small compared to the wavelength of interest; the actual way in which this is achieved is dependent upon the modeling technique in question. Mechanical CAD databases are not typically used since the information is usually extremely detailed and must be simplified for use. If very small details are included, such as bend radii, many codes will finely break up all or part of the problem to achieve full resolution of the bend. This creates an unmanageable number of unknowns. An examination of the problem geometry is necessary to reveal what level of detail should be included. Direct geometry entry through a specialized graphical user interface (GUI) with the various software implementations of the modeling techniques is typically the most efficient way to enter the model's geometry.

Dimensions less than one-tenth the size of items of interest can be ignored, especially if they are well removed from the source or area of interest. Materials located close to the source of RF energy will have disproportionally large effects, so it is important to include such details if there is any doubt as to their effects. Similarly, for remote components of the model, only details greater than one-fifth wavelength need be included. As the modeling of a problem progresses it is valuable to use various levels of complexity so that the final model can be as complete as possible.

The frequency range of interest for the problem is also known. This is important, as it determines the shortest wavelength of interest. It is useful to partition problems to have a minimum of 10 elements per wavelength at the highest frequency, although in some cases finer partitioning may be needed. Similarly it is essential that there is no excitation at frequencies where the partitioning is insufficient as this can cause instability and errors in the final result.

In order to create a model, there must be a known source. This can be a normalized source or an actual source. A normalized source level of 1 volt would typically be used as the excitation for an MoM antenna model. Once the simulation is performed, all details of the antenna parameters are known and radiated power can be easily scaled from the original 1-volt source level. An actual source is one where the source conditions are known, for example, the spectral content of a trapezoidal pulse on a PCB, or an interfering field strength used to determine immunity.

The final element of the problem that should be examined is the environment in which the device being modeled is to be located. The radiated emissions from a personal computer may be predicted from a table top, but the model might be constructed very differently if the same computer is located in a rack or even in a protective enclosure on the factory floor. All conductors in the vicinity must be considered for inclusion in the model.

6.5.4 Graphics

One of the most difficult aspects in understanding electromagnetics is the inability to visualize the fields. The ability of many electromagnetics codes to provide clear graphical plots of fields is therefore of great benefit when examining EMI/EMC problems. Through graphics, coupling paths may be identified and possible means for minimizing them can become apparent.

One of the most powerful graphic options available is unique to FDTD models. Since the FDTD technique is a time domain based technique, it has the ability to show the propagation of the source pulse throughout the computational domain. It must be recognized that the pulse is actually comprised of many frequency components and some will propagate differently than others.

A quasi-static model can be used to show distributions in space for DC and low frequency fields. Plots can show current, charge, equipotential lines, and much more.

FEM and MoM models can be used to produce field distribution plots at specific frequencies. As all field terms are available, it is possible to generate many different kinds of plots, from the basic E and H fields to everything that can be derived from them.

The MoM technique solves for the current distribution across the surfaces of the model. This basic information can be displayed in a variety of ways. One of the strengths of MoM models is the ability to calculate field values at any point in space which makes it ideal for plotting the radiation patterns of antennas, or unintentional radiators.

There are numerous applications where the main output from an EMI/EMC model needs only to be graphical. These include confirming that the model correctly represents the desired problem; checking the model during its development to both identify errors and to gain insight into the EM behavior of the structure being modeled; and demonstrating to non EM engineers how, why, and where the generated RF energy is propagating. Graphical options include simple graphs which show the field strength at a given point; field plots which show a static view of the field at one frequency or one point in time; and full motion videos of wave propagation. EMI/EMC is very much an abstract subject for many; therefore, graphics can help bridge the gap in understanding.

The examination of field plots and other intermediate results is also a powerful way to determine the correctness of a model. Section 9.4 discusses this subject in greater detail.

The EMI/EMC challenge is to understand the current flow and field behavior within and around the system of interest. Once this is done the task of controlling them becomes much less complicated. It is easier to understand the solution in a visual plot rather than the solution in the form of a set of mathematical expressions. Simply observing the behavior of a propagating pulse or the standing wave pattern in a cavity can provide insight to the experienced engineer, suggesting alternative solutions and preventing the selection of poor solutions. Areas of very high field strength can be identified and the benefits of local shielding easily evaluated. The discontinuity of the return current where an etch passes over a break in the reference plane can clearly be shown as well as evaluated for impact on radiation and signal integrity.

The task of EMI/EMC design has one side to it which is usually far from technical. Others must be convinced that the extra expense, time and testing is really necessary. Computational electromagnetics, with its ability to produce graphic outputs in the form of field plots or movies, is a powerful tool in this area just as it is for the hard core design and analysis tasks. Graphics can be very revealing as, by showing field behavior or RF current distribution, it is possible to gain insight into the EM behavior of the structure being modeled. EMI/EMC is only recently beginning to gain acknowledgment as a true science rather than black magic. The resulting increase in the credibility of the EMI/EMC design process should not be underestimated. This aspect of EMI/EMC modeling provides the designer with tools to clearly present to product managers and other non EM specialists why recommended approaches must be taken to control a particular problem.

The method by which field plots and movies are created is very much dependent upon the tool being used, while the choice of what to examine and in which fashion is not. When working with very complex geometries there is a greater possibility of having an element out of place. A series of field plots is one of the best ways of ensuring that the proper model geometry is represented. This could be a series of two-dimensional slices taken across the computational domain or a full 3D movie. An important point to consider is the memory space required to hold the graphics data, and the additional time it will take to produce it. The best guide here is to create the minimum needed except when it is to be used for a presentation where the visual quality is more important. Two-dimensional snap shots taken at widely spaced intervals will require less space than a two-dimensional movie, and much less than a full 3D movie.

Data from simple observation points can also be used to some extent to provide a deeper understanding of a system's electromagnetic characteristics. This is best achieved by using numerous observation points. Typically these will be located at and between critical points in the model. Placing observation points close to, and on both sides of apertures in an EMI/EMC shield, in addition to the source and main observation points, will provide an insight into how the energy is being reflected and how its amplitude varies with distance. This is by no means as easy to use as a good field plot or movie, but is still a major step up from no knowledge at all.

It is said that a picture can replace a thousand words. A good field plot can replace not only a thousand words but also a great deal of heavy mathematics. The creation of detailed graphics for error checking, understanding, and even presentation, should be a serious consideration for all EMI/EMC modeling tasks.

6.6 How to Approach EMI/EMC Modeling

There are two basic approaches to EMI/EMC modeling. Absolute field strength modeling uses the true source amplitude as determined by measurements or a separate simulation to predict the final emissions amplitude which can be compared to the regulatory limits. Relative field strength modeling typically uses a normalized input amplitude and compares the difference in the results as a particular feature (for example, the length of an aperture) is varied.

Both approaches are valid and are useful in different applications. While absolute field strength modeling allows comparison against regulatory limits, it requires more careful attention to the details involved when making the measurements. Section 9.3 discusses potential sources of discrepancy when comparing model results to measurement results.

Relative field strength modeling allows the engineer to create idealized and isolated models. While it is physically impossible to create either an idealized or an isolated system in the real world for measurement purposes, such models allow the user to focus on the desired effect without the need to consider other effects. Idealized and isolated models are an extremely powerful tool for the modeling engineer. The engineer can analyze specific features of a potential design without the need to create complex laboratory experiments. Often it is impossible to create a laboratory experiment to analyze a particular feature that can be very conveniently modeled.

6.6.1 Idealized Models

The use of idealized models can be very simply implemented by using plane wave or infinitesimal sources which do not play a major role in how a model behaves other than providing a controlled excitation. More complex possibilities exist that permit the extraction of detailed data from a model that would be extremely difficult to measure experimentally.

For example, consider the study of the many modes possible in a waveguide, from evanescent to high-order modes. With modeling it is possible to create a specific mode free of contamination from other modes. This provides a means of studying a complex behavior in a very well-defined and specific manner; something not possible experimentally. It must be stressed that to create a perfectly clean model, a great detail of attention must be paid to the design of the model, so that incorrect symmetries or imperfections do not cause the generation of other modes, just as would be the case for a practical experiment.

Another example of an idealized model is the filter components and decoupling capacitors on a PCB. Real-world capacitors have significant inductance which limits their usefulness as filter components at high frequencies. However, when the initial model is developed, the placement of these filter components and decoupling capacitors is also critical. An idealized model could be used to replace the capacitors with a short circuit (zero ohms) at all frequencies to simulate an ideal capacitor. Once the proper location of the ideal capacitor is determined, the ideal capacitor can be replaced with models that include the true capacitance, inductance, and resistance to determine the correct capacitance value. Obviously such ideal capacitors are not available for experimental use, but their use in EMI/EMC modeling helps reduce the amount of time needed to complete the design.

6.6.2 Isolated Models

The idea of isolating certain parts of a problem is by no means new to EMI/EMC design practices. The goal is to create a model which accurately represents a certain part of the problem at hand in a detailed manner. An example of this approach would be the careful modeling of an EMI/EMC shield where the goal is to understand the many effects of hole shape, size, spacing, and proximity to other conductors. The specific effects of the source to shield or other nearby conductors is ignored so that a good understanding of a particular mechanism can be obtained.

Another example of an isolated model is a microstrip on a ground reference plane. A model can be constructed so that a source and load is present on the structure with the microstrip. Power supply cables, ICs, and other real-world features can all be ignored. The model can then be used to study the effects of moving the microstrip closer to the edge of the ground reference plane, or the effect of different termination impedances, or the effect of voids and splits in the ground reference plane. Each can be studied separately, giving the engineer significant insight into the important causes of EMI/EMC problems.

The ability to study individual aspects of a complex problem is the greatest benefit of creating isolated models. There is a danger, however, that data obtained in this manner can be misinterpreted. In the real problem there may be other, more dominant, aspects to the problem that have been totally ignored. The experience and knowledge of the EMI/EMC engineer is needed to ensure the correct feature has been isolated and modeled.

6.7 Summary

This chapter discusses a number of important considerations when preparing to create EMI/EMC models. Modeling can be performed with either quasi-static or full wave tools, depending on the frequency range and the physical size of the device to be modeled. The various modeling techniques (FDTD, FEM, and MoM) were briefly discussed and the various strengths and weaknesses of each technique explained.

A number of different elements are required for a given model. The source type and excitation are critical and must be carefully chosen. Chapter 7 will present a step-by-step approach to creating models and provide a number of examples explaining the implementation of these.

Chapter 7

Creating EMI/EMC Models

7.1 Creating Practical Models

The art of creating practical EMI/EMC models is to ensure that only sufficient detail is included to represent the physical problem accurately. Too many details can lead to overly complex models that require excessive computer resources. Too few details and the results obtained may not provide a true picture of the problem's behavior. Fortunately the middle road between the two extremes is fairly wide, and the balancing act is one that is mostly learned from experience in modeling specific types of problems. A set of generic model construction plans are given to provide a starting point for the creation of any EMI/EMC model. These plans are then used to develop specific examples, which have been chosen to highlight the strengths of the various techniques.

7.1.1 Model Creation With FDTD

The Finite-Difference Time-Domain (FDTD) technique is perhaps one of the most versatile used in the field of electromagnetics simulation. For EMI/EMC work, FDTD modeling offers a robust method for solving Maxwell's equations in a particular region of space, the computational domain, no matter how complex the geometry or how varied the materials within that region. Further, FDTD permits a wide range of sources, observation points, and graphical options to the user. The ability to work in the time domain with FDTD is extremely powerful, but it must be used carefully in some circumstances.

FDTD is a powerful method for solving electromagnetic problems. When used to model EMI/EMC problems, it can be used to its fullest when applied to cases in which the model geometry is very detailed and when results are required for a wide frequency range. When using FDTD to create EMI/EMC models, the models should be prepared in the following order:

1. Define the required cell size for the model
2. Select the absorbing boundary conditions
3. Define the size of the computational domain
4. Enter the problem geometry and material characteristics
5. Add sources
6. Add observation points, including any field extension requirements

Cell size. Chapter 6 discusses the minimum requirements for an acceptable cell size. The general guideline requires the cell dimensions to be no more than one-tenth the shortest wavelength. Also, the physical size of model components must be considered. It is often the case that a sheet metal enclosure will have a wall thickness of less than that required for the minimum cell, as called for by the highest frequency present. In these cases the choice has to be made on whether to consider the item in question to have zero thickness, or whether it must be fully represented. This must be resolved by considering each case individually.

Zero thickness occurs when a single plane of field components is used to represent the sheet metal. For cases in which there are no openings in the sheet metal, or openings with dimensions much greater than the sheet metal thickness, it is usually acceptable to use a single cell to represent the thickness (infinitely thin or zero thickness).

To represent a specific thickness of conductor, the tangential electric fields must be set to zero on both surfaces; therefore, in general, the number of cells required to represent this thickness properly will be $n+1$. For cases in which openings are of similar dimensions to the material thickness, it may be necessary to use multiple cells to represent the thickness of material so that the fields are adequately resolved. In practice, it has been found that three cells usually provide sufficient resolution. The x-, y,- and z-dimensions of a cell do not have to be identical, so careful planning can minimize the final number of cells required. To prevent excessive numerical dispersion, FDTD models should have cell dimensions in no more than a 3:1 ratio between the cell dimensions.

Absorbing boundary conditions. There is a wide range of possible absorbing boundary conditions available to truncate an FDTD computational domain. However, within any particular tool, there may only be a limited selection available. Some boundary conditions offer high absorption for direct illumination and others offer compromises over a wide range of incident angles. For specific, well-defined problems, it is possible to minimize the size of the computational domain by selecting the conditions that best match the problem. This, in turn, means shorter solution times with lower computer resources. The selection should be based on a knowledge of the expected direction of propagation of the fields within the computational domain.

Computational domain. The FDTD computational domain is the volume throughout which the simulation will be run. It is required to contain the model and enough white space around the model elements so that the absorbing boundary conditions can function properly. The final dimensions will be set by the number of cells required to detail each of the x, y, and z-axes. It is important to remember that the computational domain is measured in cell units and not by the physical dimensions of the problem. The use of very finc cclls for a given physical space will greatly increase the size of the computational domain.

Problem geometry. The problem geometry is the heart of the FDTD model. This step is completed by setting each of the cells of the computational domain to have appropriate material properties. Each cell can have any electrical properties, including free space, perfect electric or magnetic conductivity, or any combination of parameters. Other cells will be set to represent the source. The mechanics of entering the cell data is usually buried in a Graphical User Interface (GUI), so that objects are placed into the domain together with their properties. Thus it is not necessary for the user to literally enter data for each cell.

Source. Once the main problem geometry has been created, it is necessary to add the source and its excitation waveform. There are many considerations when defining the source for an FDTD model. The source can be defined by a plane wave, an impressed current in space, a voltage or electric field between two conductors, a current on a conductor, or a magnetic field. The plane wave, which is very common for scattering applications, is ideal for immunity models where the primary problem is to determine how much coupling results from an external field, which fortunately is usually specified as a plane wave. The impressed current source, while it is a nonphysical source, can be extremely useful. It has no self-resonances, nor can it couple to other components of the model, because it is mathematically defined within the computational domain with no physical conductor to support it. Voltage and current sources are circuit model sources and are very similar in that they both have an amplitude and source impedance term. Electric and magnetic field sources are also in common use; these are usually specified between and around conductors, respectively. The challenge to the user is to select the model source which most accurately describes the EMI/EMC source of the problem.

The source excitation for the FDTD model can either be added into the field update equations, or it can be forced. If it is added, the excitation will die down and the fields in the region of the source can continue unperturbed (this is known as soft excitation). If the excitation is forced, the field update equations are essentially defined, with the result of specifying the fields around the source for the duration of the simulation (known as hard excitation). For most EMI/EMC work, the former method of source implementation is the best.

A series of sources can be used to force particular modes onto a structure. Although often this does not truly represent a practical source, it can be a powerful diagnostic tool to help understand a particular behavior. Using such a series of sources would be an example of an idealized model.

The unique aspect of the FDTD technique is that it solves in the time domain. This provides the possibility for obtaining output data over a wide frequency range through the use of the Fourier transforms. However, the frequency range is not infinite and, if energy is present above or below the valid range of the model, errors will result. The excitation wave form must be selected to ensure that its spectral content is high over the range of interest but minimal elsewhere. Depending upon the tool in use, this waveform will be specified by the maximum and minimum frequency required, or by the pulse characteristics directly.

Observation points. The final step in creating the FDTD model is to add the observation points and field extension techniques. The observation points will be used to monitor the field behavior both within the computational domain and, through the use of field extension techniques, beyond it.

To ensure that the data required is obtained from an observation point, this point must be set up properly. Within the computational domain, field observation points and circuit probes can be clearly identified with respect to other geometry details and the results can therefore be more easily understood. However, they can still be subject to misinterpretation. For example, if only the tangential electric field is monitored in close proximity to a good conductor, the levels will be low, while the magnetic field and radial electric field components will be much higher.

It is important to consider the characteristics of the observation points. Each observation point acts as a perfect linearly polarized, isotropic receive antenna. These points provide a measure of the true value for the field at a given point within the computational domain. This is of particular importance when considering correlation to a measured result obtained from an antenna or other probe which will have some directivity and not be perfectly polarized.

To ensure that the far-field observation points are used correctly, it is essential to know whether they provide a full wave extrapolation or whether they assume that the observation point lies in the far-field. It is quite possible that both choices are available. The most probable

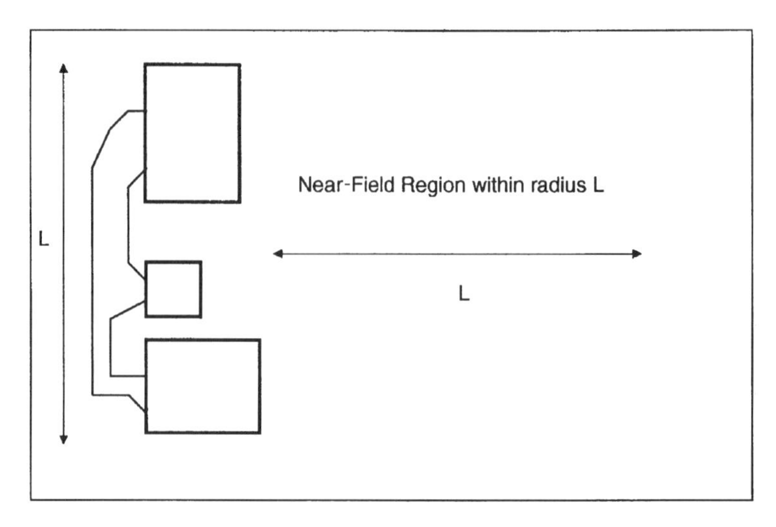

Figure 7.1 Near-Field Region for a Large System

error that can occur would be the use of an extension technique which calculates only the far-field values for a location that is still in the near-field of the radiator. In such a case, the dominant near-field contributions could be ignored. For a dipole, the change from near-field to far-field regions occurs at approximately one-sixth of a wavelength; however, for a large system, this does not hold true. As a guide, the near-field region of any array of conductors that comprise the system should be considered to extend radially from that system for a dimension equal to the largest dimension of the system. This is shown in Figure 7.1.

7.1.2 Practical Considerations for FDTD Modeling

Many EMI/EMC problems are controlled by means of locating the radiating components, together with many other conductors, within a shielded enclosure. When excited, any conductor will ring at its natural resonant frequency with an amplitude which decreases depending upon the load or dampening present. In the idealized world often used for modeling, conductors are perfect and so the excitation of a wire or two parallel walls can result in very high Q resonances.

These resonances are important to consider, as they can cause extended ringing when modeling and play a critical role in determining the coupling within an enclosure. The FDTD technique, in particular, is vulnerable to ringing of this nature as, once a field is set to propagate between loss free conductors, within the perfectly aligned domain of a model there is no inherent mechanism by which the fields can decay. However, the model can be constructed in a number of ways to minimize this ringing without affecting the accuracy of the final data.

The simplest approach in avoiding ringing related to resonances is to make the structure asymmetric and lossy. The structure should be prevented from being completely enclosed, if at all possible, by leaving at least one side open to the absorbing boundary. This stems from the basic rule of modeling—include only the elements that are really needed. This is not always a possibility, so other avenues of model construction must be explored.

The problem of ringing lies in the loss free nature of the materials used. Introducing losses will cause the model to reach a steady state, although experience has shown that the losses required are larger than those expected simply from using real conductivities. Another option is to avoid perfect alignment of the enclosure walls. Real enclosures have some degree of bowing in the side walls or a draft angle which ensures that perfect alignment cannot occur. Providing there is some loss in the model, adding even a small angle to an FDTD model will cause the ringing to decay rapidly.

An example of the effects of ringing on various angled walls can be seen in Figures 7.2 through 7.7. Figure 7.2 presents a diagrammatic illustration of a simple problem of a source wire being excited and an observation point where the field strength must be determined. Four FDTD model cases were created, each with a slightly different geometry.

Case 1 Source and observation point alone

Case 2 Source and observation point in the presence of two parallel walls

Case 3 Source and observation point with one wall at a slight angle

Case 4 Source and observation point with one wall at a greater angle

Figures 7.3 through 7.6 show the time-domain responses at the observation point for each of the four cases. Figure 7.3 shows the pulse for case 1 as it propagates past the observation point with no reflections present. In case 2, with two parallel walls added, Figure 7.4 shows

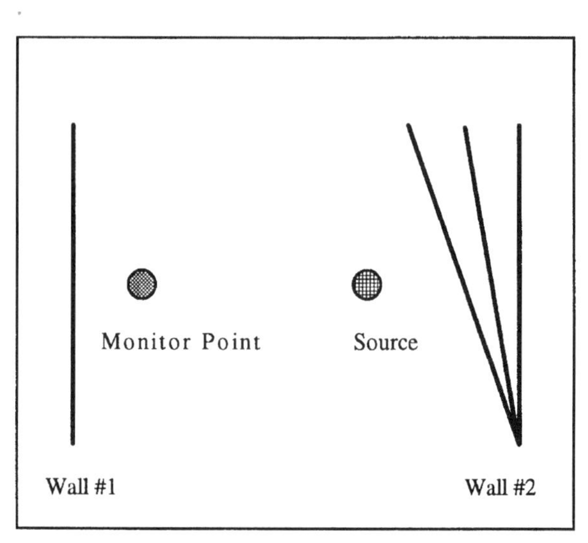

Figure 7.2 Source and Monitor Point With Angled Walls

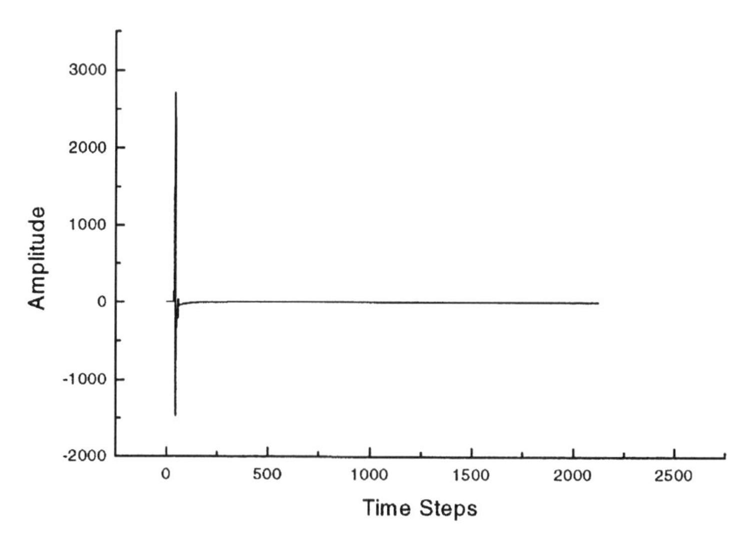

Figure 7.3 Time-Domain Response With No Walls

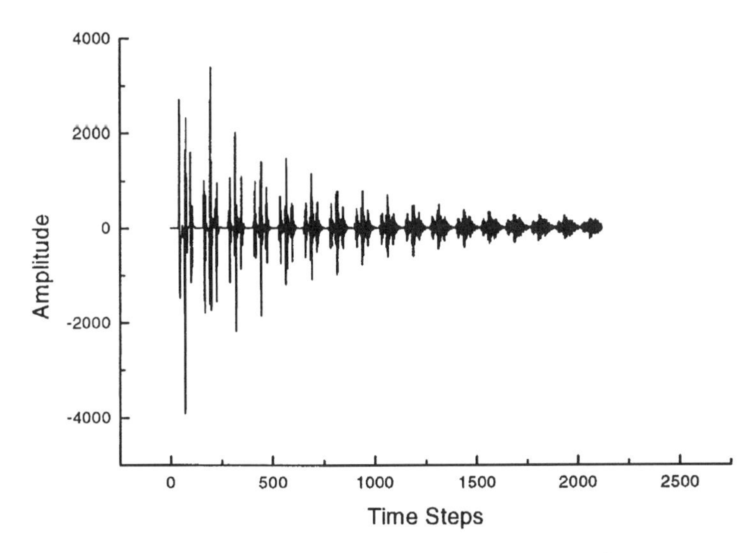

Figure 7.4 Time-Domain Response With Parallel Walls

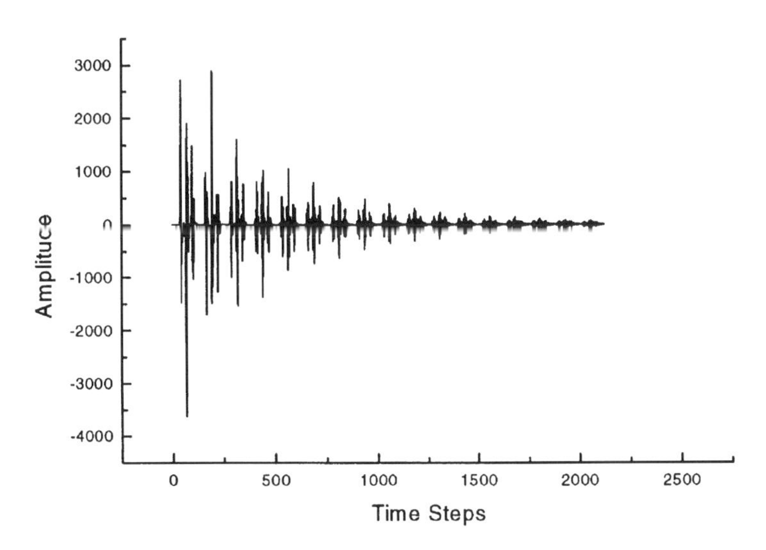

Figure 7.5 Time-Domain Response With Small Angle on One Wall

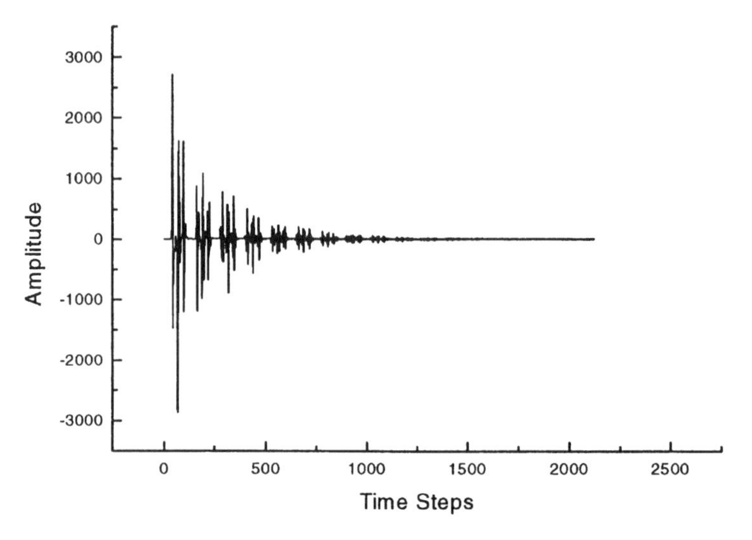

Figure 7.6 Time-Domain Response for Steeper Wall Angle

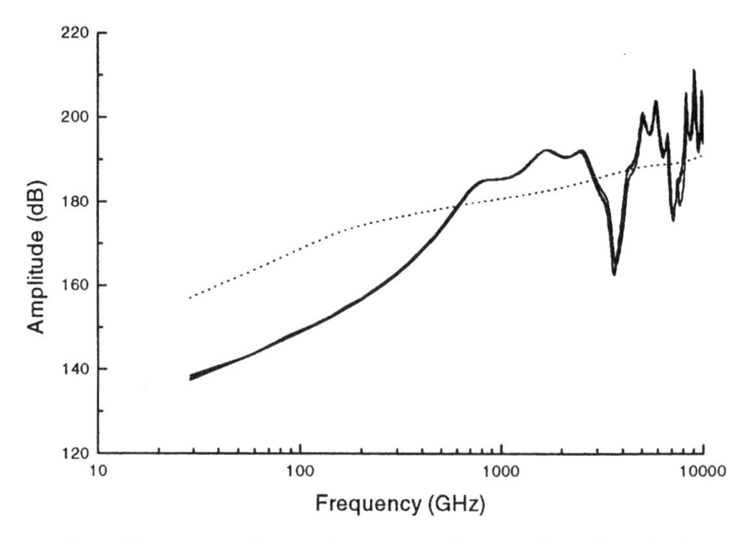

Figure 7.7 Frequncy-Domain Response for All Cases (No walls dashed.)

that multiple high-amplitude reflections are observed. Figures 7.5 and 7.6 show how much faster the fields settle down in cases 3 and 4, with one of the walls inclined at two different, yet small angles. For this to be a valid method for controlling extended ringing, it is important that the frequency response characteristics are not significantly changed by the introduction of the angled walls.

Figure 7.7 shows the frequency response for all four cases. Clearly, the walls cannot be simply excluded from the model, as there are major differences between the first and second cases, with and without walls. However, it is clearly shown that the frequency responses of the parallel and inclined wall cases are very similar.

Eliminating perfect alignment by using inclined walls is a powerful modeling technique, permitting fields to decay more rapidly in enclosed areas. The degree of frequency domain distortion is generally quite small; however, it is prudent to examine the implications of these distortions with reference to the modeling task at hand.

Provided there is some decay to the ringing there is the possibility of extrapolating the ringing wave form mathematically. Application of the technique known as Prony's method is a powerful tool that can greatly shorten solution run time for resonant structures.

7.1.3 Model Creation With FEM

In general, the Finite Element Method (FEM) can be used for the same classes of problems as FDTD, with the restriction that only results for one frequency can be obtained with each run. Operating in the frequency domain makes FEM models less sensitive than FDTD models to resonance conditions. However, this can be considered both a disadvantage and an advantage. It is often very desirable to know the frequency at which a device resonates; using the FEM technique it is necessary to use fine frequency steps to avoid missing a critical frequency. On the positive side there is very little chance of instability at frequencies close to resonance unless the model is excited at the precise resonant frequency.

When using FEM for creating EMI/EMC models, the models should be prepared in the following order:

1. Enter the problem geometry and material properties.
2. Add sources.
3. Add observation points.
4. Select the absorbing boundary conditions.
5. Mesh the computational space.

Problem geometry. The EMI/EMC problem's geometry and associated material properties can be entered with the use of a GUI that will take care of much of the housekeeping. An advantage for FEM models is that it is possible to use the same simplified models created for structural analysis, as these are usually free from the small details that are not required for EMI/EMC modeling. This can make it much easier to obtain suitable databases, minimizing the amount of data re-entry needed.

Sources. The same types of source are used in FEM as FDTD: a plane wave, impressed current in space, voltage or electric field between two conductors, current on a conductor, or magnetic field. However, as the source is defined in the frequency domain, it is less complex than that required for FDTD. The source must be selected to best represent the problem at hand. FEM models are very common for signal integrity analysis (parameter extraction) and some are well suited to interface with circuit simulation tools. This integration makes the task of specifying circuit probes and sources very easy.

Observation points. The selection and placement of observation points have the same requirements as for an FDTD model. Field observation points may be located both within the computational domain and external to it, and the same care must be taken to ensure that the wanted data is obtained. With FEM, observation points that are external to the computational domain are calculated from a surface that completely surrounds the model elements. This surface should be located as close as convenient to the model elements.

Absorbing boundary conditions. In FEM, the absorbing boundary conditions only have to deal with one frequency at a time; however, the more complex meshing usually used with this technique results in higher reflection levels from the boundary than possible in FDTD. Therefore, it is necessary to have a greater amount of white space between the FEM model and the absorbing boundary. Most FEM Absorbing Boundary Conditions (ABCs) require a circular or spherical computational domain, adding to the white space required. The white space requirement will be dependent upon the wavelength of the frequency of interest. Fortunately, in FEM, the mesh granularity is not fixed but can become relatively coarse, perhaps as large as one-fifth of a wavelength, especially in the white space area away from any small model details.

Meshing. Meshing an FEM model is an extremely complex issue. The mesh is usually created automatically through built in meshing sub routines. It is common that the mesh will be automatically refined to ensure sufficient resolution in areas of small detail. However, the user often has the ability to fine tune the resulting mesh by selecting areas of specific concern and forcing the creation of a finer mesh in those areas.

7.1.4 Practical Considerations for FEM Modeling

The meshing techniques for FEM provide an excellent resolution of complex geometries, as there is no fixed element size. Further, in the case of non penetrable objects such as a perfect electrical conductor, only the surface of the object needs to be meshed, not the entire volume. Depending on the mix of elements in the model, this can result in a significant reduction in the complexity of the model.

7.1.5 Model Creation With MoM

The Method of Moments (MoM) is a versatile technique that has many uses for EMI/EMC modeling. MoM models can be created for both radiating systems and parameter extraction, although these frequently require different implementations of the technique.

The basic method used to prepare an MoM model follows the same outline no matter what the purpose of the model, and is listed below.

1. Enter the problem geometry, including the presence of ground planes.
2. Define the frequency range.
3. Define meshing of the elements (if necessary).
4. Add the sources, loads, and materials.
5. Add the observation points.

Problem geometry. The first step in creating an EMI/EMC model using the MoM technique consists of defining and entering the problem geometry. Surfaces may be modeled as surface patches or wire frames. In general a wire frame model can be solved in a shorter time than its surface patch equivalent, at a small expense in accuracy. For radiation models, wire frame models are usually more than adequate, while for the quasi-static implementation, surface patches are generally used to ensure accuracy.

When modeling with surface patches, the geometry data can be entered by using a GUI and the computer left to take care of the meshing at a later stage in the process. However, if a wire frame is to be constructed, it is essential to create the framework with sufficient detail to permit an accurate representation of the current distribution. At one extreme, for example, a complex box, requires meshing, the guideline being a mesh density of one tenth wavelength for the highest frequency of interest. At the other extreme, long wires may be used to represent attached cables and similar model elements. All long wires must again be meshed finely to ensure that the current distribution along it can be fully represented.

There are additional constraints which apply to wire elements. The diameter of the wire is usually considered to be extremely small compared to the wavelength of interest such that no current can flow circumferentially. Also, the segment length must be longer than the wire radius.

The pattern used for framing an object can be optimized if something is known about the source locations. Currents on a plane will travel radially from a feed point so a radial pattern of wires is a good choice for surfaces when a feed wire is attached as shown in Figures 7.8 and 7.9. Figure 7.8 shows a feed wire attachment to a small surface, while Figure 7.9 shows how the feed wire attaches to a larger surface. The number of radials should be sufficient such that their far ends are no

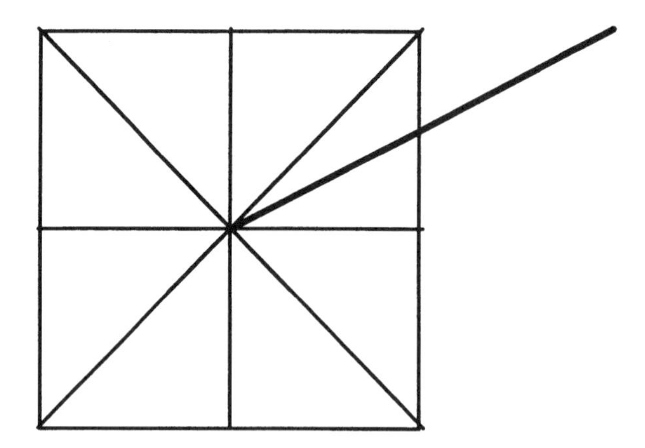

Figure 7.8 Attaching a Wire to a Small Surface

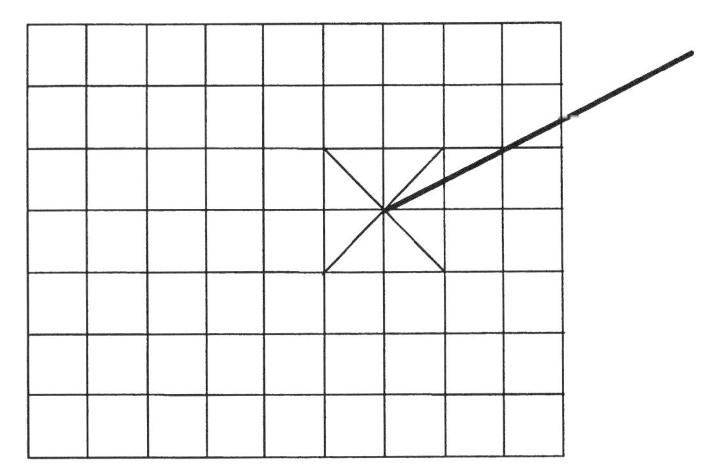

Figure 7.9 Attaching a Wire to a Large Surface

more than approximately one-tenth of a wavelength apart, which should match with the overall meshing scheme. This permits for modeling circumferential currents on the edge of the surface. For more distant surfaces, when no knowledge of the current pattern can be assumed, a cross-hatch pattern as shown in Figure 7.10 is usually best. The goal is to produce a model that will support the real current distribution with the minimum number of wires to keep the computational requirements reasonable. Details of the problem geometry that are less than one-tenth wavelength for the highest frequency of interest may often be

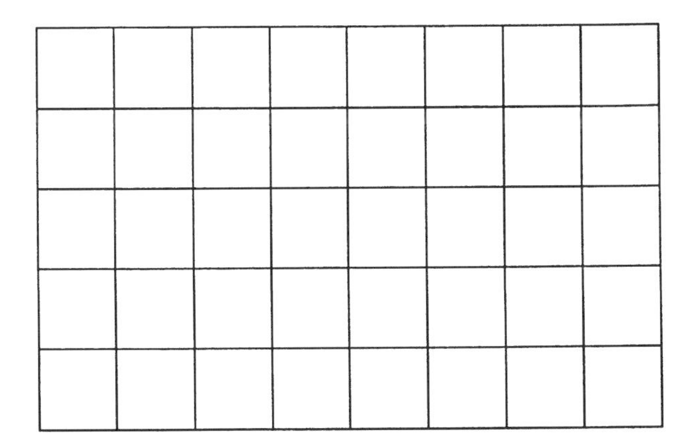

Figure 7.10 Cross-Hatch Wire Frame Pattern

ignored providing that the overall problem geometry is also not very small.

The presence of a ground plane is very common in EMI/EMC measurements and is an important detail that must be included in practical models. The ground plane is readily implemented in the MoM technique by using symmetry and does not add to the solution time significantly.

Frequency range. It is important to know the complete range of frequencies for which an MoM model is to be solved. The meshing will be defined by the highest frequency of interest; therefore, it is possible to use the same model for perhaps as much as two decades below this frequency. This is often enough for a practical EMI/EMC problem of 30 MHz to 3 GHz; however, wider frequency ranges will require more than one version of the model to best match the required resolution against the solution time.

Meshing. The meshing (commonly known as segmentation within the MoM user community) of the problem elements may be done manually or automatically. This is totally dependent upon the specific tool being used and how the technique is implemented. As MoM codes solve in the frequency domain, it may be desirable to create models with different mesh densities, as oftentimes a wide frequency range is needed to optimize the solution time and memory requirements. In all cases the requirement is to ensure that all individual elements are meshed to a minimum of one tenth of a wavelength for the highest frequency of interest.

Sources. The next step in creating an MoM model is to add the sources, loads, and materials. Adding the source or sources is often done by replacing one or more small model elements with a voltage or current source. Since the MoM technique solves for the current distribution on all the segments, it is very easy to include lumped circuit values within the model whether they are sources or loads (R, L, and C).

The inclusion of materials such as dielectrics is more complex in MoM than in either FDTD or FEM. This is because the effects of such materials are based not only on their surface properties, but on their volume as well. Therefore, there is no easy mechanism for including their effects within a standard MoM model. Special codes are available that model the displacement current on the dielectric's surface-to-air interface, but these are typically complex to use and require expert users. Some other specific cases are addressed, such as lossy (imperfect) ground planes; these can be used for EMI/EMC models in a variety of ways. However, if there is a need to model various materials, the MoM technique might not be the best choice.

Observation points. The addition of observation points is the final step in creating a MoM model. MoM codes have no specifically defined computational space, so there are no truncation issues, and observation points can be placed freely. Circuit probes are also readily available as the MoM tool is solving for the current or charge distribution on all the model elements. Thus, in addition to field observation points it is possible to obtain the current in any segment of the model. This permits plots of the current distribution to be obtained which are also a useful form of output data.

A further benefit of many MoM codes is that inherent in the calculations to produce the current or charge distribution is the coupling between elements. This results in the ability to easily obtain both mutual and self inductance or capacitance between elements of the model.

MoM codes may provide preset field observation point selections. It is very important to ensure that these are appropriate for use with the given EMI/EMC problem. For example, commercial radiated emission limits require that radiated emissions be measured over the surface of a cylinder with a given radius, usually 3 or 10 m, and over a height range of 1 to 4 m. While some codes have these locations as a preselected choice they may also offer a "maximum field strength" over a hemisphere that can be much more severe and will also minimize the possibility of matching simulation results to measurements. See Section 9.3 for further information on this subject.

7.1.6 Practical Considerations for MoM Modeling

The primary consideration when creating a MoM model is that it is a surface-based technique. However, in most of the practical tools based on the MoM, the surface is single sided. This means that, for example, if a model of a shield is created, the same currents will be flowing on both the inside and outside of the shield. This is not a true representation of how an EMI/EMC shield works, however.

The limitations that result from having the same currents on both sides of the shield can be avoided in two ways. First, two simulations may be run. One simulation would solve for the current distribution on the inside of the shield, providing a source value to be used in the second simulation. The source for the second simulation would typically be the fields in an aperture or set of apertures. The second simulation would solve for the current distribution on the outside of the system, given the source found with the first simulation. The second way of solving the shielding current problem is to use a specially developed MoM code that permits independent currents on either side of the surface. While such codes do exist, they are uncommon at present and very complex mathematically.

Parameter extraction models using the MoM technique will often employ the quasi-static approximation to simplify the problem and may solve for current or charge distributions, depending on the application. This is an excellent way to shorten calculation time when all of the problem dimensions are small compared to the wavelength at the frequency of interest.

7.2 Modeling Electromagnetic Radiators

Electromagnetic radiation is a function of current flow on *all* the exterior surfaces of the device or system under consideration. This is the one of the most important points to consider when developing a computational model. It should be remembered that the MoM techniques are surface-based techniques, so there is a natural match between MoM and radiator modeling. While an EMC engineer is concerned with complex radiating structures, these are really antennas that lack a clear or obvious geometry. Considerable resources are available for antenna design and analysis work that can be applied to this kind of EMI/EMC problem.

To simulate the emissions from a system accurately, it is necessary to reproduce the radiating structure accurately in the model being developed. This must include all conductors, i.e., ground straps, interface cables, and enclosures, which may influence the current distribution and hence the total radiation. Emission peaks occur when the source energy is most closely matched to the antenna presented by the structure being excited. Resonances that relate to multiples of approximately half-wavelength dipoles are typically, although not always, the worst cases.

While controlling radiated emissions is one of the greater challenges in the field of EMC, it is also necessary to ensure that systems have acceptable levels of immunity to external electromagnetic sources. Using reciprocity, a system that does not radiate effectively will also have greater immunity. Therefore, an accurate radiation model can be used to provide an equally accurate immunity model; a given excitation permits the calculation of the field strength at a given location. Once this relationship is known in relative terms, the energy induced at the source by a given external field can be calculated. This is very useful, as most EMI/EMC modeling codes emphasize emissions rather than immunity analysis.

7.2.1 Modeling a 30-MHz Half-Wave Dipole

Commercial EMI regulations require measuring the field strength radiated from a device under very specific conditions. The distance between the antenna and device is set, typically to 3 or 10 m. The antenna is scanned in height from 1 to 4 m, using both horizontal and vertical polarizations for the antenna. The device itself is arranged in such a way as to maximize the emissions. Finally the frequency range covered extends from 30 MHz to at least 1 GHz.

Most laboratory radio frequency (RF) measurements, such as insertion loss and phase response, are precise and repeatable to within a small fraction of a decibel. Typical open area test site measurements, as required for regulatory compliance, are neither high-precision nor easily repeatable. One area that sets the underlying accuracy of radiated emission measurements is the antenna used.

The half-wave dipole is the antenna specified by the Federal Communications Commission (FCC) to be used as the official reference for making field strength measurements. At the lower frequency portion of the measurement range there are a number of problems associated with such a physically large antenna. One aspect of the dipole's behavior that can be examined is the antenna input impedance. For communication applications, antennas are properly matched to the feed cable; this assumes that the antenna impedance is fixed. However, as the antenna height and polarization are varied above the test site ground plane, this impedance changes.

An example is detailed below of the use of numerical modeling to determine the behavior of a 30-MHz half-wave dipole. The goal is to establish a value for the measurement uncertainty when using this antenna. Using the procedures set forth in Chapter 6, this problem is defined, and the modeling and computational techniques chosen to best fit the goals.

Given an ideal, thin, 30-MHz, half-wave dipole in free space, it is known that the input impedance at resonance is 73 ohms. A practical antenna with finite thickness elements will have a slightly lower impedance. To determine how a practical 30-MHz dipole behaves, the input impedance must be examined for both polarizations and compared to the free space value.

When creating a model of the dipole geometry, three essential details must be included: the element diameter, element length, and antenna's location with respect to the ground plane. Other details may or may not be required; these include the variations in diameter that result from using telescopic elements, the droop in the elements due to their weight, slight misalignments from the horizontal or vertical, and slight errors in setting the element lengths equally. One other element which may be included is the feed cable. A key factor to consider during vertically polarized measurements is that the drape of the feed cable can significantly impact the antenna response. For this example, only the essential details are examined, but the same procedure used here is appropriate for covering all the additional details.

As stated at the beginning of this section, the use of the MoM technique is a natural choice for antenna design and analysis, and is the best choice for this 30-MHz dipole problem. With this problem defined and the goal of obtaining the input impedance of the 30-MHz dipole established, the final steps as outlined in Section 7.1 can be followed. These steps are:

1. Enter the problem geometry.
2. Define the frequency range.
3. Define meshing of the elements.
4. Add the sources, loads, and materials.
5. Add the observation points.

For this problem only, a single frequency, 30 MHz, is entered in the model. Each of the 30-MHz dipole elements is given to have a length of 2.413 m, measured from the center of the balun, and to have an average diameter of 6 mm. These numbers are taken directly from the tuning chart that is shipped with a generic dipole set. The free space antenna factor is given as –1.8 dB. It should also be noted that the balun is designed to match the free space antenna impedance to 50 ohms, and any change in the antenna impedance represents a mismatch loss.

The dipole is modeled as a single long wire having a length equal to the total length of the dipole (4.826 m) and a diameter of 6 mm. The source is located at the center of the wire. The source selection is unimportant when calculating impedance. This is because the impedance is a fixed value for a given configuration and, no matter what the voltage or current used to excite the model, the ratio between voltage and current is always defined by the antenna impedance. For simplicity a 1-volt source is used. There is no need for field observation points in this model because, in the process of solving for the currents, the input impedance is obtained as well.

Antenna factor relates the voltage at the antenna terminal to the electric field in which it is immersed. Most standards call for the antenna to be calibrated in a "free space" environment. A free space model is used to provide a reference for the other configurations. To simulate free space, the first model is run without the presence of a ground plane; this means that there are no other conductors in the model to affect the antenna behavior in any way. The antenna geometry is shown in Figure 7.11, and the segmentation used is shown in Figure 7.12.

Owing to the effects of the ground plane, practical experience has shown that at 30 MHz, for the 3- and 10-m test distances normally used, emissions will peak when the antenna is at its highest point due to the destructive interference between the source and its image.

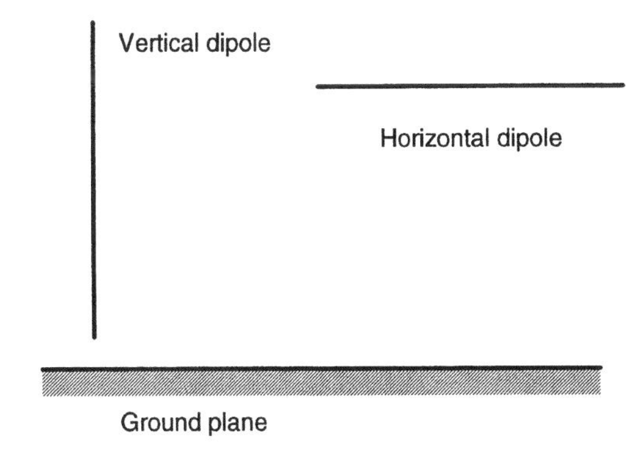

Figure 7.11 30-MHz Dipole, Vertical and Horizontal Above a Ground Plane

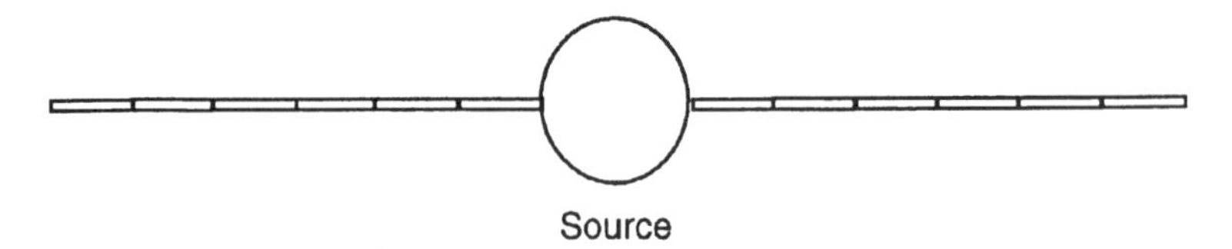

Figure 7.12 Model of Dipole Showing Segmentation

Therefore, the second model has the 30-MHz dipole positioned horizontally at 4 m above a perfectly conducting infinite ground plane.

For the vertical case, the fields are strongest in the plane of the ground plane, as the interference is constructive. Hence, the third model has the dipole positioned vertically with its lower tip 250 mm above the perfect ground plane.

For all three models, the wire is meshed into 13 segments, with the source located on the center segment. As this is a half-wavelength dipole, there are 26 elements per wavelength that will provide a good solution without requiring excessive computational time.

The results obtained for the input impedance of the dipole and the associated mismatch loss, assuming perfect match to the free space case, are shown in Table 7.1. The results of this modeling show that even under ideal conditions the use of a half wavelength dipole at 30 MHz can introduce an error of approximately 1 dB into the measured results. It should be noted that half-wave dipoles sent out for calibration have been returned with antenna factors of close to 0 dB at 30 MHz, rather than the theoretical −1.8 dB. This is to a large part explained by the mismatch loss associated with the calibration, where the antenna is positioned horizontally at 4 m above the ground plane.

It is possible to extend this model to include the presence of the feed cable and even a source antenna for site attenuation measurements. In these models, the mutual coupling between antennas is automatically

Table 7.1 30-MHz Dipole Properties for Different Environments

Condition	Resistive Value (ohm)	Reactive Value (ohm)	Mismatch Loss (dB)
Free space	71.0	+j 0.26	0.00
Horizontal	87.4	−j 13.00	0.95
Vertical	93.8	+j 2.10	1.29

included, a parameter that has been widely neglected until now. While source impedance is a critical factor in determining the mismatch loss, the influence of all other conductors on the antenna radiation pattern may also be considerable. New site attenuation models are being developed using the MoM to provide more accurate data than previously available, by including all the critical aspects of the site into one model.

To illustrate the importance of using modeling to understand these site related issues, the effect of the feed cable location on the behavior of the 30-MHz dipole is examined for the vertically polarized case. The dipole model used is the same as for the vertical case above, with the sole addition of a second vertical wire used to represent the feed cable. The feed cable is often set back from the antenna by a distance of 1 to 2 m and is often passed through the ground plane with a feedthrough connector that bonds the coaxial shield to the ground plane. The feed cable extends from the ground plane to the center of the dipole, at a height of 2.663 m. The resulting model is shown in Figure 7.13. The feed cable is initially positioned at a distance of one meter from the antenna and then moved to 2 m in one-tenth-meter increments. Table 7.2 lists the antenna input impedance and the mismatch loss caused by the change in impedance from the free space case.

It can be clearly seen from these data that significant measurement errors can result if the feed cable is too close to the antenna. This has two consequences. First, if the antenna feed is in a fixed geometry, there could be a systematic error. Second, if the feed is free to drape

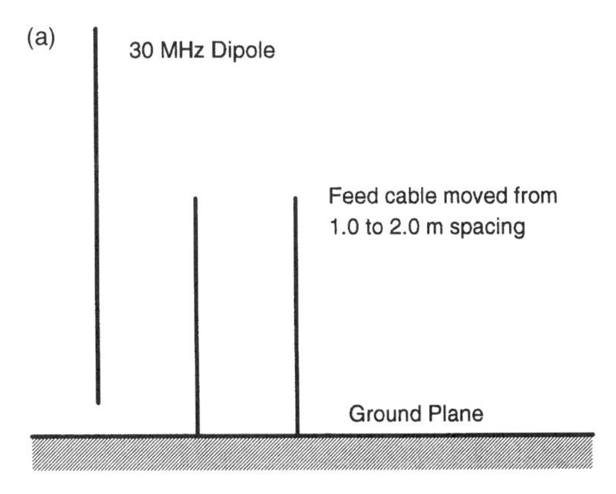

Figure 7.13 Model for Vertical Dipole With Feed Cable

as it may, an additional random error may well be introduced. Most important to note is the magnitude of the error, 2.3 dB, created by position changes of only 0.5 m.

A commonly stated goal is to have EMI measurements repeatable within 2 dB. As the variation in the value of the mismatch loss obtained by modeling is so high, it seems appropriate that this type of model should be extended to consider the many practical site considerations still unexplored that can limit measurement accuracy and repeatability. Other considerations include the real feed cable drape, how the cable is terminated at the ground plane, the effect of excess feed cable lying on the ground plane, and probably many more.

7.2.2 Modeling Real Systems as Dipoles

Many types of systems under test for emissions limits resemble a dipole-like structure. That is, there are one or more long wires which are oriented to permit maximum radiation, with a common-mode source voltage between at least one of the wires and part of the metal structure. For example, in the case of a shielded enclosure with a number of wires and cables exiting from the real panel area, a common-mode

Table 7.2 Results for 30-MHz Vertical Dipole With Different Feed Locations

Feed Location (m)	Resistance (Ohm)	Reactance (Ohm)	Mismatch Loss (dB)
Antenna alone	93.8	+j 0.26	1.29
1.0	37.8	+j 1.16	2.31
1.1	44.3	+j 8.80	1.78
1.2	50.6	+j 14.9	1.28
1.3	56.7	+j 14.9	0.82
1.4	62.7	+j 23.0	0.40
1.5	68.3	+j 25.4	0.02
1.6	73.7	+j 26.9	0.31
1.7	78.7	+j 27.0	0.60
1.8	83.3	+j 27.9	0.86
1.9	87.6	+j 27.5	1.09
2.0	91.4	+j 26.6	1.28

j = symbol for $\sqrt{-1}$; it indicates an imaginary number.

voltage between one of the cables and the shielded enclosure would create a dipole-like structure.

When the shielded enclosure is electrically very small, it could be approximated by a single wire, resulting in a very simple, non-center-fed dipole-like antenna. However, when the size of the enclosure is not small, the enclosure dimensions must be included in the model.

For this example, we assume that we know the common-mode voltage at the cable connector from a previous model, or from measurements. Therefore, the model is only an "outside" model; that is, there is no attempt to model the different fields inside and outside the enclosure. Only the outside fields are of interest.

We must first select the modeling technique appropriate for this problem. Since long wires are to be attached to the shielded enclosure and no shielding effectiveness modeling is needed, the MoM technique is the technique best suited to this problem. Both the FDTD and the FEM techniques would require excessive computer RAM to model the long wire structures.

Models of long wires have been covered in the previous section. However, for this problem, we must model the outside structure of the shielding enclosure as well. While the MoM technique allows either surface patches or wire segments, we will use a wire frame model of the enclosure for this application.[1]

The most common approach to wire frame structure modeling uses a rectangular grid of wires to cover the structure everywhere except where a long wire attaches to that structure. The wire grid size must be small compared to the wavelength (1/10th λ normally used), and the individual wire segments must connect to other segments only at their ends. Where other wires are attached to the structure, the wire grid is modified to be a radial pattern so that the RF currents are allowed to travel in any direction necessary. The long wire with the common-mode voltage source is then attached to the center of the radial wire grid.

For example, consider a shielded box with dimensions of 0.5 m $\times$ 0.3 m $\times$ 0.2 m. If the frequency range of interest for this model was from 100 MHz to 1 GHz, the shortest wavelength is about 0.33 m. Therefore the wire frame grid size should be no more than 0.03 m, resulting in a mesh of 17 $\times$ 10 $\times$ 7 grids. Figure 7.14 shows an example

[1]Some implementations of surface patches in the MoM technique do not permit accurate modeling of long wires attached to a surface patch.

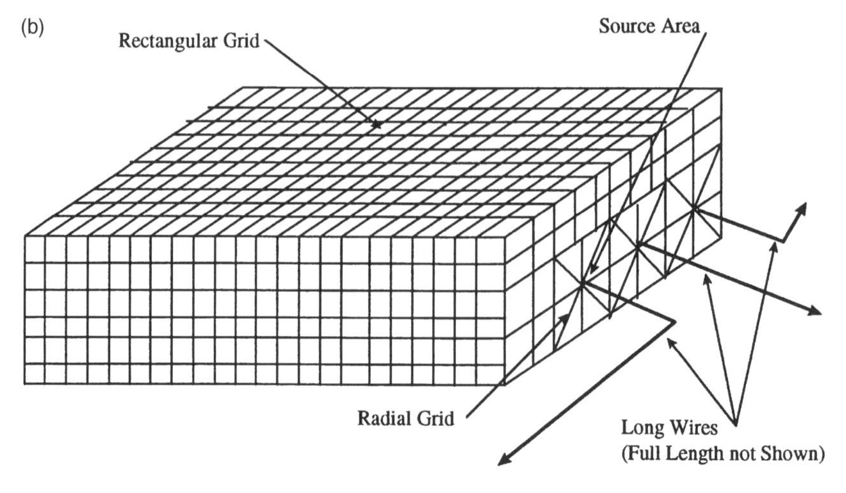

Figure 7.14

of this model with three long wires attaching to the side of the enclosure. One wire contains the common-mode source voltage. This source is placed on a short wire segment nearest the enclosure.

Once the model geometry is completed, the field monitor points must be specified. This structure is fairly complicated and will not have a simple dipole-like radiation pattern. Since it is convenient to specify far-field monitor points in the MoM technique, we will specify the monitor points must cover a region at 10-m distance from the model, from 1 to 4 m high, and in all directions (as if the model were rotated on a turntable). The maximum frequency of interest is 1 GHz, with a wavelength of 0.3 m. At 10 m, the total circumference around the receive locations is about 60 m. Therefore, we must ensure the field points are measured at a sufficiently small distance from each other to guarantee the highest field points will be measured. For EMI applications, 1-degree measurements (360 points) have been seen to be precise enough for the required accuracy. Obviously, more points can be added if a higher resolution is desired. Likewise, the vertical scan height of 1 to 4 m should have about 30 points for the same resolution. Typically for EMI applications, only the highest amplitude of all the 10,800 spatial points are kept.

The general example given in this section can be modified in a number of ways. For example, the position of the cables can be changed. They can be connected to another shielded box, or terminated to the ground plane. Additional external long wires can be added and one or more of them used as the source wire.

During this example, all wires have been considered to be directly bonded to the shielded box. However, it is quite possible that a cable will have a ferrite core over it with the intent to block common mode currents. This does not mean that the cable can be completely ignored. The only real difference is that the end condition of such a cable is open, not shorted. Very often, the cable is still closely coupled to other cables and the shielded box and so may be a factor in the system emissions. To include a ferrite choked cable in the MoM model, it is simply entered into the model as an isolated wire of the appropriate length and location. It is prudent to consider the real effect of the ferrite choke to the model since these components are not perfect. One way in which this is done is to obtain the input impedance from the feed point for the cable when it is driven with respect to the box. For the cable to be considered open, the ferrite impedance must be much larger than the antenna impedance for the range of frequencies being considered.

As an example of how modeling can greatly benefit the EMI/EMC engineer during the engineering phase of product development, a method is given to make the most of bench top voltage measurements and minimize the need for the more complex radiated emission measurements. It is possible to create many different models, each with a possible system layout. The goal is to develop a model that represents all variations of the cable position and box placement. Once done and the maximum field strength for a given source is known (often 1 volt), it is easy to use the ratio of voltage to maximum field strength to determine the maximum allowable voltage while keeping emissions below the required limits. This permits the EMI/EMC engineer to focus on controlling the common-mode voltage at the connector directly.

7.2.3 Heat Sink Models

In today's high-performance computers, clock frequencies extend well above 100 MHz. Additionally, the power dissipated by these devices has also increased such that heat sinks of significant size are required to ensure reliable operation over the expected ambient temperature range. These heat sinks are closely coupled to the RF energy source, due to the fast, high current switching that takes place in such processors.

Therefore radiation from the heat sink of a Very Large Scale Integrated (VLSI) device can be a significant source of EMI. Figure 7.15 shows the structure of the VLSI device and its heat-sink.

When creating a model to examine the effects that a heat sink will have on emission levels, it is important to consider what is really needed. In a computer using fast processors, it is common to use a shielded enclosure. Therefore, the goal for this task is to develop a model that can be used not only alone, as a radiator, but also as a true representation of an EMI source in a more detailed model.

Radiation modeling requires a full-wave technique and a full three-dimensional model to be created to represent the problem geometry. A further consideration is that a wide frequency range must be covered, typically over a decade, as many harmonics are generated by high-speed switching circuitry. The geometry of a heat sink over a module's ground plane has the potential to have many resonant frequencies, therefore a time domain is the ideal choice for this problem. The use of a frequency domain technique would require many passes to ensure that the full frequency response is obtained, and the use of either the MoM or FEM technique would become extremely repetitive and time-consuming. FDTD is therefore the logical technique to select for this work, as it provides a detailed spectral response and permits the creation of a small model to investigate the fundamentals that can be easily integrated into a more complex model.

The problem geometry, including the dimensions of the heat sink, source model, and ground plane, is shown in Figure 7.16. The heat sink fins are included in the overall dimensions of the heat sink as the

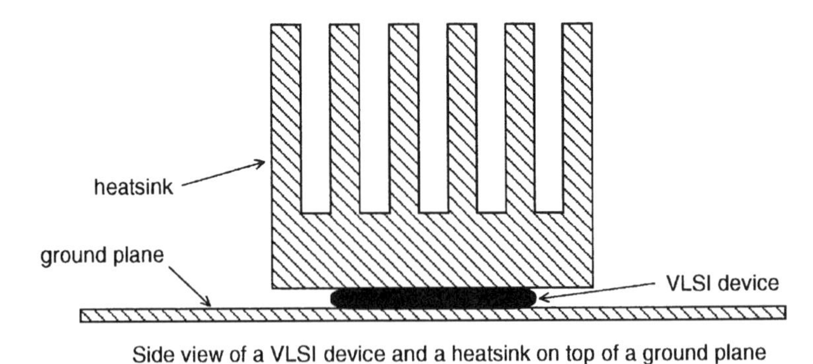

Figure 7.15 Heat Sink Problem

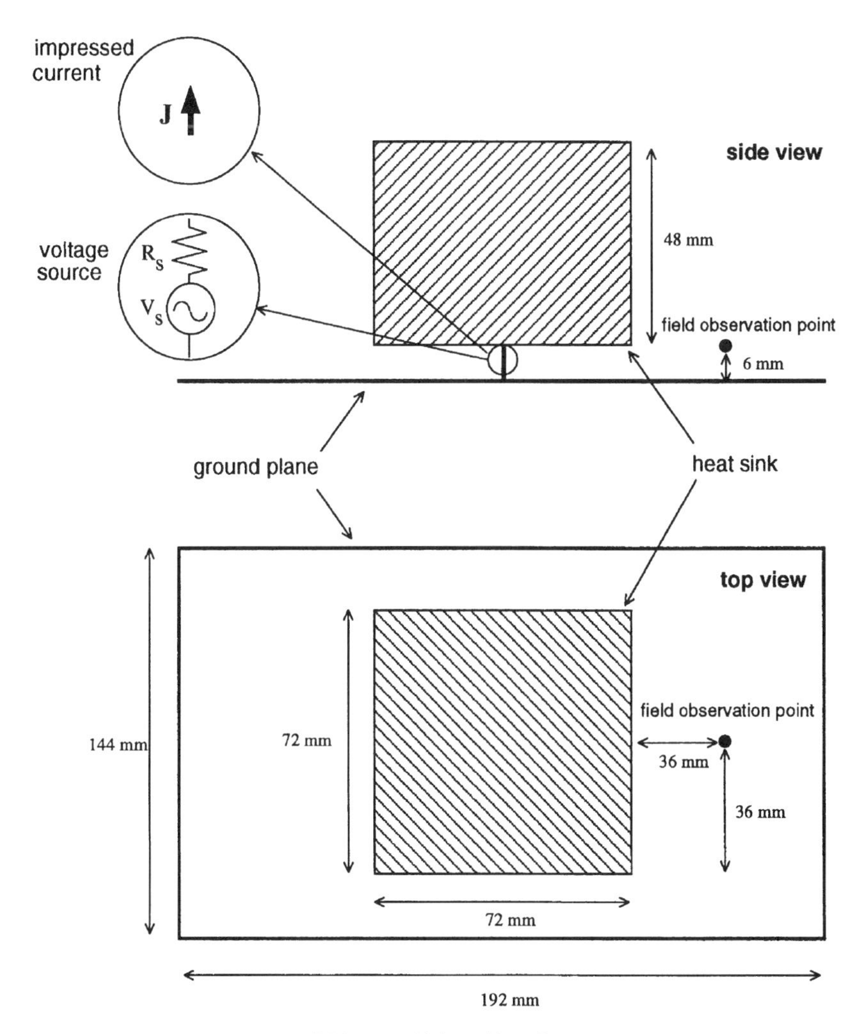

Figure 7.16 Heat Sink Problem Geometry

gaps between the fins are small compared to the detail of interest. The source model represents the VLSI device package. The frequency range of interest for this problem is from 100 MHz to at least 3 GHz.

In general, all geometry details of less than one-tenth of a wavelength for the highest frequency of interest will have only a minimal effect. Therefore, they need not be included in the model geometry. The maximum frequency of interest for this example problem is 3 GHz, and the minimum dimensional detail of interest is about 10 mm. Some geometry details, however, are less than 10 mm but will be of interest because they are located close to the source. This includes the source model height (gap between the heat sink and ground plane).

Now that the problem has been defined, goals established, problem geometry detailed, and modeling and computational techniques chosen, the steps listed in Section 7.1 may be followed to create this heat sink model.

Cell size. The first step is to define the required cell size for the model. A cell size of 3 mm is selected for the x- and y-dimensions; this is the maximum size that can be used and is acceptable as there are no small details in these orientations. In contrast there is detail in the z-direction, which must be fully modeled; this is the gap between the heat sink and the ground plane. To ensure adequate resolution, a cell size of 1.5 mm is selected; this permits four cells in the gap. These cell size selections permit the use of the smallest computational domain.

Absorbing boundary conditions. When selecting the absorbing boundary conditions to be used for this model, a goal was to minimize the computational resources required. The complementary absorbing boundary condition is chosen for this model as it provides the lowest reflection levels when close to the model elements. This enables the computational domain to be significantly reduced in size as only a minimum of white space is needed.

The computational domain has two regions, the model region and the white space region. The model dimensions are 54 mm high × 144 mm wide × 192 mm long, as shown in Figure 7.16. The corresponding cell dimensions are 1.5 mm, 3 mm, and 3 mm, as determined in the first step. This makes a model region of 36 cells high × 48 cells × 64 cells. The white space region required around this model is approximately 20 cells which is to be added all around the model region. The combination of maximum cell size and complementary boundary conditions permits the computational domain to be only 72 × 88 × 104 cells.

Problem geometry. The next task is to enter the problem geometry and material characteristics. The heat sink including fins is modeled as a single block having the same overall dimensions as a heat sink with fins but without the gaps between the fins. This can be done because the fins are very closely spaced, less than the 10 mm of interest.

A wire is located at the center of the heat sink to connect to the source model. The final part of the problem geometry is the ground plane. The heat sink, wire and the ground plane are modeled as perfect electrical conductors.

The case for the source model geometry is more complex. The radiation peak of the system is dependent on the exact physical dimensions, therefore this source model cannot simply be ignored. The most important feature of the computational model is that its overall dimensions match the real problem, a 6-mm VLSI device package, and can be modeled as a voltage source with appropriate source impedance in a 6-mm gap. This ensures that the heat sink is in the correct relationship to the module for calculating the emissions. The validity of these assumptions can be easily tested by creating models with different gaps or with the addition of a number of heat sink fins. Such experimentation is strongly recommended as it is an important part of the learning process for those beginning to use modeling.

Source. It is necessary to add a source that is representative of the VLSI device. For this example, the VLSI device is modeled as a resistive voltage source. This resistive voltage source is a single cell entity and connects between the ground plane and the feed wire connected to the heat sink. The resistive voltage source comprises a perfect voltage source in series with a specified resistance in a single cell. When selecting values for the source voltage, it is best to use 1 volt and then scale the results obtained to the real values. In this way, the model results are effectively normalized and can be used for different devices without modification.

Selection of the source resistance is not so simple, and it is unlikely that the impedance value will be readily available. However, it is possible to make an estimate of the source impedance. Consider that a large VLSI device may be switching tens of amps in 1 nsec or less, while maintaining the supply voltage to within about 100 mV of its nominal value. For these conditions to be met (100 mV and a current change of 10 A) the impedance value could not exceed 10 mΩ. Provided that the load impedance on this source has a significantly higher value than that of the source, it is acceptable to use a perfect voltage source with a small resistance to represent the VLSI device in the model. Fortunately, the model itself can be used to determine the load impedance of the heat sink on the VLSI device by evaluating the current that flows from the source, as is discussed in the last step, observation points. The source used in this model comprises a one volt perfect voltage source in series with a one ohm resistor.

Observation Points. The final step in creating this heat sink model is the addition of observation points. In this model there is a circuit probe, a near-field observation point, and a far-field observation point, the latter requiring the use of a field extension technique. The current probe is located around the source wire and is used in conjunction with the source voltage to determine the feed impedance of the heat sink as a radiator. As the observation points provide time domain data, it is necessary to perform a Fourier transform to obtain the frequency data. Once the voltage and current into the antenna are known, the impedance is easily calculated from Ohm's law.

Two field points are used; one is located just 36 mm from the heat sink, a location where other system components might well be mounted, and is very much in the near-field region of the heat sink. The second observation point is located in the far-field at 3 m, to provide a measure of the radiated fields at a typical test distance. The far-field surface used to calculate the 3-m point was located just two cells outside the model region. All of the data presented is for the dominant polarization, which is orthogonal to the plane of the ground plane.

Results. The results obtained from an FDTD simulation are in the form of a time domain plot. The data obtained for the near-field observation point is shown in Figure 7.17. This plot is extremely useful as it is immediately possible to determine whether the simulation has been run for sufficient time steps. As this plot shows, the time response to the excitation comprises some ringing which then dies down to zero. If the energy had not died down, errors would have resulted when the transformation to the frequency domain was performed. As the user becomes more experienced, other information can be extracted from the time domain plot, such as recognizing reflections from the absorbing boundary conditions; however, this is very dependent on the type of model being created. In general, the user will require the frequency-domain response obtained by Fourier transform from the time-domain data.

The frequency domain plots for the near and far-field observation points of this heat sink model are shown in Figures 7.18a and 7.18b.

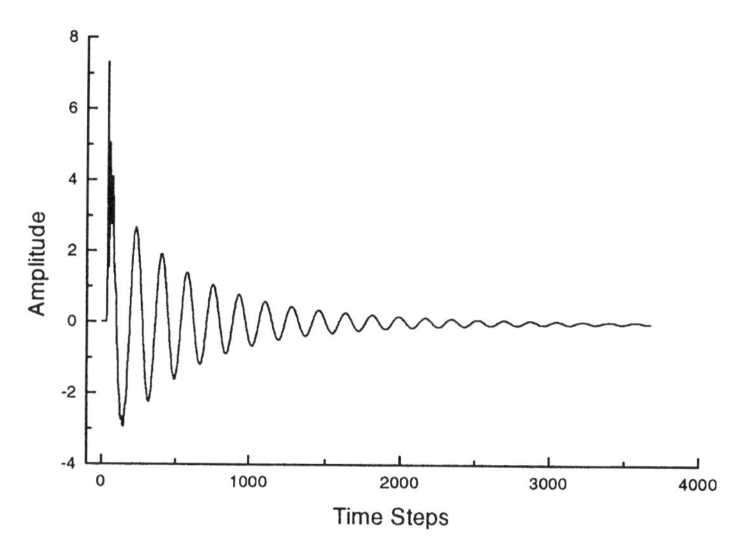

Figure 7.17 Near-Field Time-Domain Response

Figure 7.18a shows the field strength in the near-field, Figure 7.18b shows the field strength at the 3-m test distance. There are clearly differences between the frequency responses shown in Figures 7.18a and 7.18b. This is due to the additional contributions seen in the near-field from the higher-order fields (second and third order), while at the

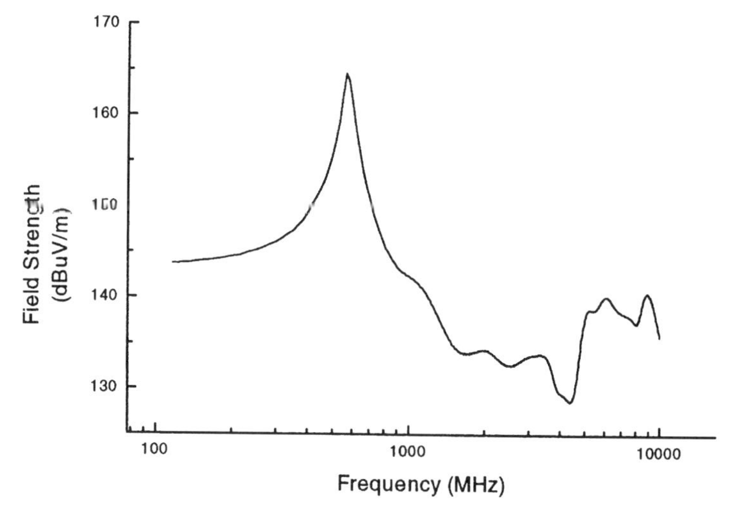

Figure 7.18a Near-Field Observation Point

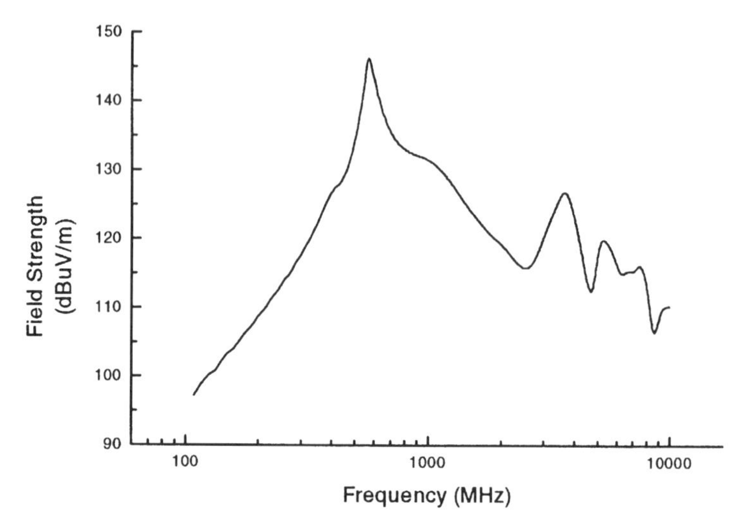

Figure 7.18b Far-Field Observation Point

3-m distance, only the radiation field is significant. The major use for the near-field data is to determine what coupling may take place to nearby circuit components, while the 3-m data represents the field strength expected if the device were not enclosed in a shielded enclosure of any kind.

Figures 7.19 and 7.20 show the real and imaginary components of

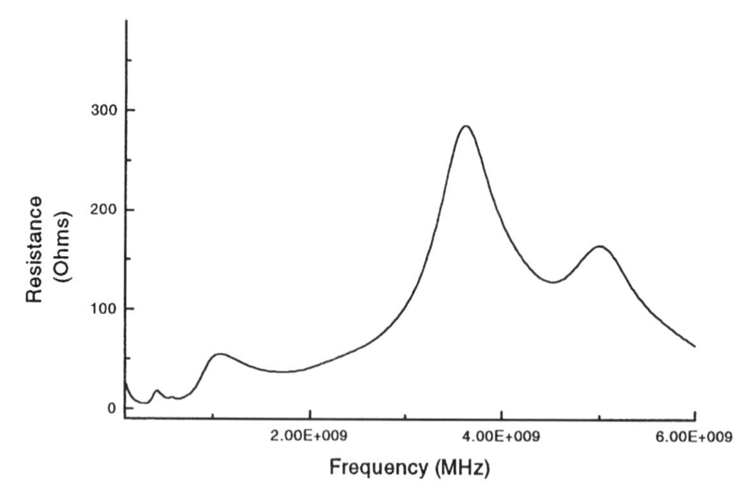

Figure 7.19 Heat Sink Radiation Resistance

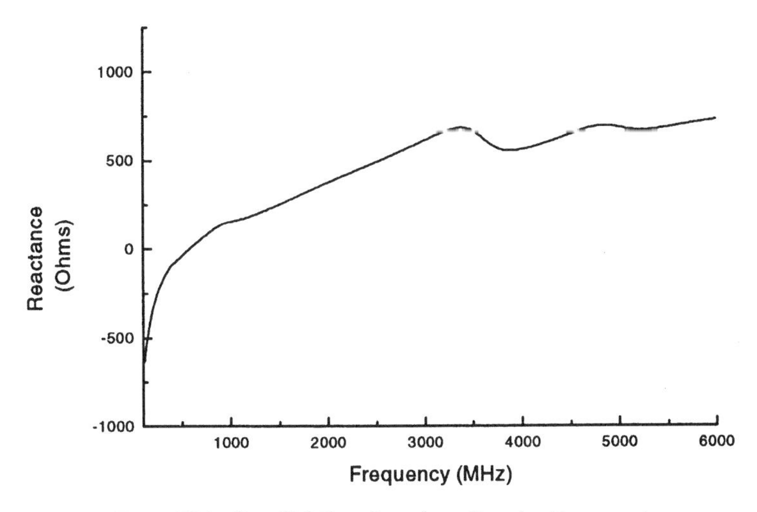

Figure 7.20 Heat Sink Input Impedance-Reactive Component

the heat sink antenna impedance, as obtained from the current probe and voltage source. It was required that the heat sink impedance be greater than 1 ohm for the source model to be considered valid. From Figure 7.19, it is clear that this condition is met across the full frequency range of interest. This validates the selection of the source model used.

This model has provided a great deal of information. The time domain response at one of the locations, was used to verify that the model run time was sufficient. The near-field and far-field components of the field were obtained at the locations of interest. Finally, the input impedance was derived to validate the initial choice of excitation. Furthermore, the model is sufficiently small so that it can be readily used as a more sophisticated source for a larger model. More detail is given in Section 7.3.

Further work. To illustrate the importance of the source resistance, the model is run a number of times, each with a different value for the source resistance. Values of 1, 10, 50, and 1000 Ω are used. For investigative purposes, it has been common practice to use very small impressed current sources to excite structures and determine how they behave. While this often has some use, it is not always as useful as expected. This is because the infinitesimal sources are unaffected by the structures they excite; unlike the cases usually encountered in practical EMI/EMC problems.

1. Heat sink with 1-ohm voltage source (true voltage source)
2. Heat sink with a 10-ohm voltage source
3. Heat sink with 50-ohm voltage source
4. Heat sink with a 1,000-ohm voltage source (current source)

The results of the simulations for both the near-field and the 3 m far-field observation points are shown in Figure 7.21 and Figure 7.22, respectively. The most apparent differences between the near-field and far-field responses occur in the low frequency regions. This is due to the significant additional contributions of the static and inductive fields to the radiation field in the immediate vicinity of the radiator.

It can be seen that for each source, both the amplitude and frequency responses are quite different. For this particular geometry, the emission peaks are much more pronounced with the lower impedance source.

While complete as a radiator model, this particular FDTD model is also one that could be used in a much larger model. Such a model could include the enclosure with its apertures, other modules, and conductors. In the model of the heat sink, it was shown how important it is to properly model the source impedance; in a larger model with other elements the source impedance is much more complex. To properly create a macro source model it is necessary to include the full

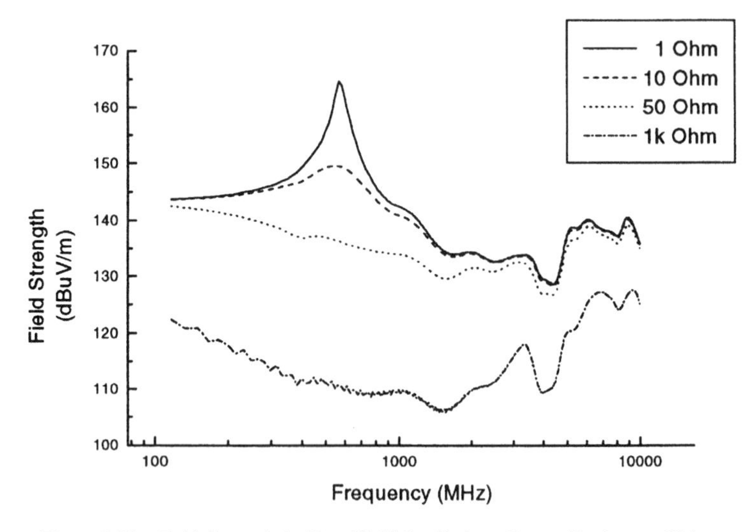

Figure 7.21 Field Strength in Near-Field for Various Source Resistance Values

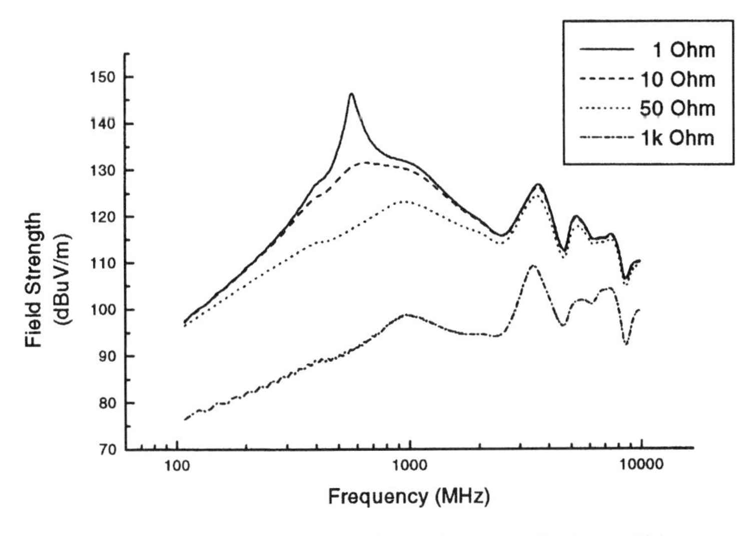

Figure 7.22 Field Strength at 3 m for Various source Resistance Values

physical representation of the source so that all the interactions between the source conductors and other elements of the model are included. In this manner a very complex model can be constructed and all the interactions between the elements of the model would be included.

7.3 Modeling a Shield With Apertures

Shielding is one of the staples of EMI/EMC control. The purpose of shielding in an electronic device is to prevent EMI from entering into, or exiting from that device. A perfect shield, composed of an unbroken conductive material, would require little analysis. Real shields, however, have imperfect seams due to the manufacturing process, and numerous openings for ventilation purposes, access panels, and connector ports. It is the presence of these seams and apertures that limits the overall performance of the shield and adds complexity to the EMI/EMC problem.

The levels of shielding used in consumer goods which must comply with commercial EMI/EMC regulations are relatively low, often less than 20 dB and seldom more than 30 dB. By comparison, a military communication system which must comply with military standards could employ shielding of greater than 80 dB. This is a critical piece of information to have when evaluating the degree of accuracy possible. When trying to achieve 80 dB of suppression, aiming for 90 ± 10 dB is probably a reasonable goal; however, trying to achieve 20 dB by aiming for 30 ± 10 dB is not reasonable. This is due to the sensitivity of the consumer product to cost and the differences in the design practices for the two very different markets. For the consumer market, one of the most important benefits of modeling is to determine correlation between EMI/EMC performance and what components or practices are used. This can result in the creation of a large number of models, each with small but well-specified differences.

Much of the shielding designed today is based on the well-known shielding effectiveness equations that, while accurate in their intended use, are frequently used for predictions that have no real relationship to their original intent. Shielding effectiveness is a specific term expressed in decibels (dB) and relates the amplitude of a plane wave impinging upon a shield to the amplitude of that plane wave after passing through. The assumptions for this are:

1. The source of the wave is remote from the shield
2. The measurement point is remote from the shield
3. The wave is a plane wave (TEM)
4. It is normally incident to the shield

These conditions imply that the shield is uniformly illuminated and that there are no other conductors in close proximity to the shield as shown in Figure 7.23. However, in many practical EMI/EMC designs, these basic requirements are not met. An example of a real shielding case is shown in Figure 7.24a. Both the source and observation points of interest are close to the shield such that plane wave conditions do not exist, and normal incidence is unlikely. Therefore, in these cases, the use of equations to estimate shielding effectiveness will not produce accurate results. To prevent costly over design it is necessary to seek out more precise methods of predicting the true performance of an EMI/EMC shield in its true electromagnetic environment.

A good example of a shielding problem is defined and solved below. The problem is approached using the principles set out in Chapter 6. Once the technique is selected, the creation of the model follows the outline set forth in Section 7.1.

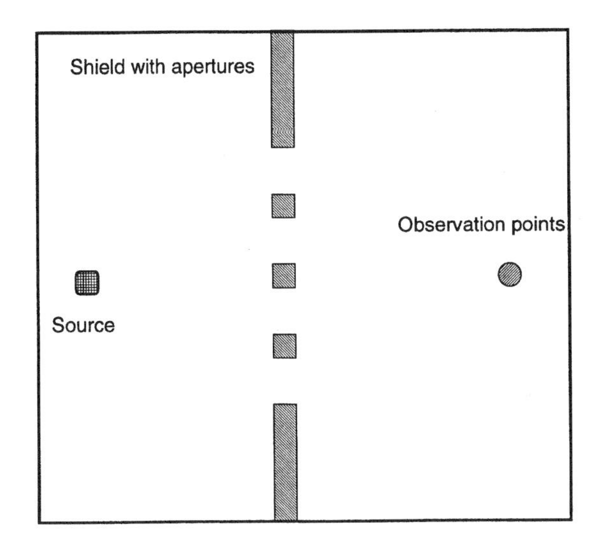

Figure 7.23 Idealized Shielding Case

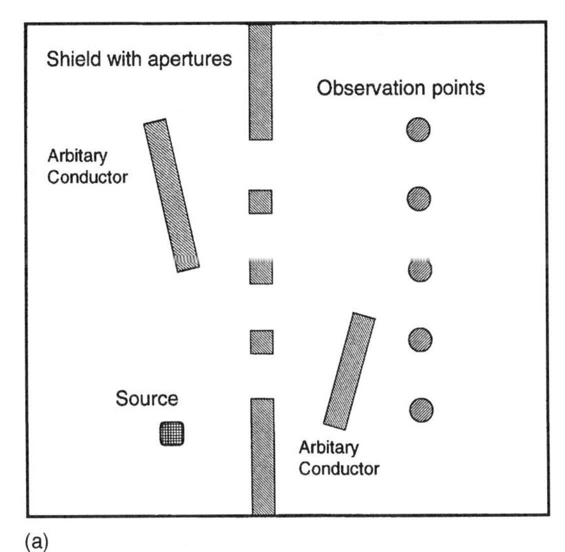

Figure 7.24a Practical Shield Problem

Given that the performance of a shield is limited by its apertures, it is often necessary to study the shielding performance of these apertures under a variety of conditions. Of particular interest are the effects of the apertures alone and in the presence of conductors both internal and external to the shield. The steps to defining this problem are:

1. Define the goals.
2. Select the modeling technique.
3. Select the computational technique.

The first step is to define the problem in terms of what is wanted from the model and what is known at the start. In this example, the goal is to obtain an understanding of the shielding performance of an array of apertures, both with and without the presence of other conductors in the vicinity of the apertures. For this problem it is required that all details of the physical geometry are known, together with the frequency range of interest. The array of apertures, each 3 mm $\times$ 15 mm in a 1-mm-thick shield, is examined for the frequency range of 500 MHz to 3 GHz. As the apertures will dominate the shield behavior, it is acceptable to consider the shield to be composed of a perfect electrical conductor for this frequency range. Further, unless the apertures are located in close proximity to the edge of the shield, the effects of the bulk of the shield may be neglected unless a detailed radiation pattern is required. Figure 7.24b shows the full problem geometry.

The selection of the most suitable modeling technique and its method of application (the second step) is based on all the above information. As the problem requires output data that are frequency dependent, a quasi-static tool will not suffice; a full wave code must be used. Further, shielding is very much a three dimensional problem rather than two-dimensional. So the final tool must be both three-dimensional and full wave. To choose between a frequency-domain and a time-domain code requires a closer inspection of the problem and a consideration of what additions may be made to the model before the work is completed. If the problem is to be a simple planar shield with a remote source and remote observation point, there may be little to choose between the two domains. However, with the addition of the internal and external conductors, it can be expected that resonances will occur that can be difficult to detect if the problem is solved only on a frequency by frequency basis. A further consideration is that it is necessary to fully

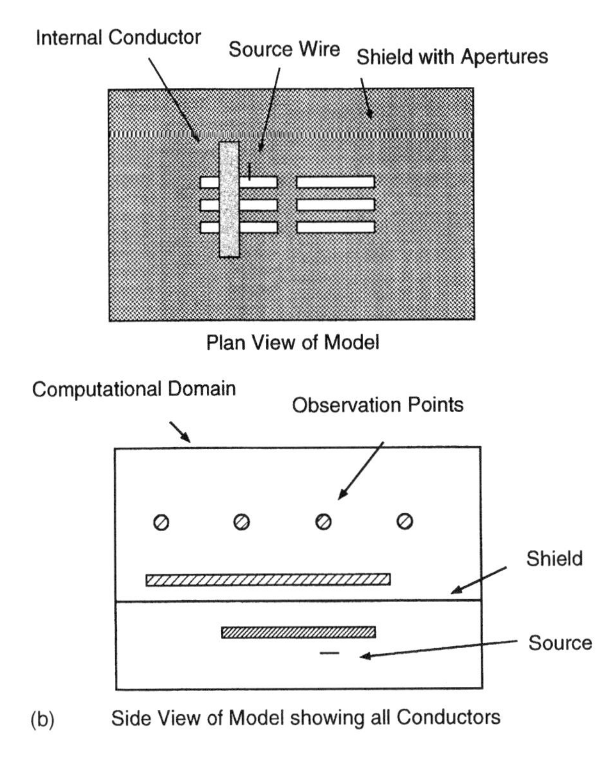

Figure 7.24b Shielding Problem Geometry

model the currents on both sides of the shield. From this discussion it is clear that the best selection for the computational technique to be used for this problem is the FDTD technique as it can provide a full spectrum of data over the frequency range required.

Having set the groundwork, it is possible to begin to create an FDTD model of the shield following the procedure set out in Section 7.1.

Cell size. The selection of the cell dimensions is based upon the highest frequency of interest and the smallest dimension of interest. The highest frequency required for inclusion in this model is 3 GHz. To ensure sufficient excitation over the required frequency range, the source pulse used will, of necessity, have energy in excess of this. To determine the maximum frequency component in the excitation, an understanding of the driving wave form is required; this will be made available within the preprocessor for a particular tool or must be obtained from the spectrum of the source pulse. For this example, the energy extends to 10 GHz, which corresponds to a wavelength of 30 mm. The frequency imposed minimum cell size is one-tenth of this wavelength, i.e. 3 mm or less.

The second cell size criteria is the smallest geometry detail of interest. The material of the shield is 1 mm thick, and it must be determined whether this has to be fully represented in the model or can be considered to be zero thickness. This determination is made based on the aperture width and height and a little general shielding knowledge. Where high levels of shielding are required, honeycomb materials are often used. These work as the apertures behave as a series of wave guides beyond the cutoff points. When the shield thickness is approximately one-tenth of the aperture's longest dimension, a 3-dB improvement in attenuation is possible; therefore, the actual thickness should be included in the model. Where the apertures are longer than this, they can be safely considered to be of one cell or zero thickness.

For this example, the apertures are 15 mm × 3 mm in the 1-mm-thick material. As the 1 mm thickness is less than one-tenth the largest dimension, 15 mm, it is possible to use a single cell to represent the shield thickness without adding error to the solution. While there is no need to resolve the fields in the thickness of the shield, it is necessary to resolve them along the 3-mm dimension, requiring a minimum of three cells. Therefore, cell dimensions of 1 mm are required to model both the 3-mm and 1-mm dimensions of the aperture. The cell dimensions for the 15-mm dimension could be as large as 3 mm, which would satisfy the highest-frequency conditions and provide an adequate resolution of the fields. However, it is also necessary to consider what other elements are to be added to the model during the study, such as source and external conductors. To ensure adequate flexibility in the model, a 1-mm cubic cell is used. This decision requires three times as much computer memory as would be the case for 1 × 1 × 3-mm cells. This is an important consideration since there is a direct trade off between accuracy and computer resources, both memory and processor time.

Absorbing boundary conditions. In this shielding example, the focus of the model is to study coupling through the apertures under a controlled set of conditions. To aid this work, it is necessary to eliminate any unwanted influences; a task that can be less complicated when modeling, than when trying to make measurements. The selection of the absorbing boundary condition has a key role in how the model can be constructed. Some boundary conditions such as Higdon and Liao use only field points that are orthogonal to the boundary of the computational domain, while others such as Mur use both orthogonal and tangential points. A major advantage is gained through the use of the former class of absorbing boundary conditions for this problem; the shield can be positioned completely across the computational domain without the need for an enclosure around the source. In this way, there are no cavity resonances or other unwanted coupling effects with which it would be necessary to contend.

Computational domain. In order to determine the size of the computational domain it is necessary to consider the size of the shield and the space required around it for the other model elements, including the white space for the absorbing boundary conditions. The apertures under study are 3 mm × 15 mm and are arranged as was shown in Figure 7.24b. As the apertures are located away from the edge of the shield, the x- and y-dimensions for the computational domain can be fixed. The third axis must be sufficient to permit the expected source, observation points, and both internal and external conductors to fit within the domain, again allowing sufficient white space around the full model. In this example, the computational domain is 51 × 72 × 80 cells.

In conjunction with the size of the computational domain, the number of time steps required for the simulation can be set. The minimum number is set by the time it takes the pulse to propagate across the computational domain. Additional time steps are required to allow any ringing due to slot resonances, etc., to dampen to an acceptable level. The total number of time steps indirectly sets the minimum frequency for the data as the overall simulation time maps to the lowest frequency in the spectrum of the output data. A total of 4,000 time steps were used for this simulation when no additional conductors were present. When conductors were added the number of time steps was extended to 9,100, to permit the increased ringing to decay.

Problem geometry. As the goal of this shielding model is to compare the shielding performance for a number of different problem geometries, several simulations must be run. In this example, five cases are examined below.

1. The first model serves as a reference and consists of only the source and five observation points; with no EMI shielding.
2. The shield with apertures is then added between the source and the observation points
3. To the third model is added a conductor in close proximity to the source and apertures.
4. In the fourth model the conductor is removed and a different conductor is added into the model on the outside of the EMI shield and in close proximity to the apertures.
5. The fifth model incorporates both additional internal and external conductors.

Source. The primary goal for this work is to study the shielding performance of a set of apertures both with and without the presence of other conductors, but without influences from the source. To avoid direct influences from the source, two good choices are available, the impressed current source or a real current source on a very short wire. The impressed current source has no physical existence in the computational domain, which makes it the ideal choice. Similarly, a very short (single-cell) wire has very little mutual impedance with any other element in the model, making it an acceptable choice. For this example, the latter source was chosen with a wire length of one cell. This was excited with a Gaussian pulse.

Observation points. Observation points must be placed in the region of interest; and for this example, a number are placed close to the shield apertures. The highest field value calculated for all points at each frequency was used as the measure of energy that penetrated the shield. Using multiple observation points, the effects of the radiation pattern are minimized.

Analysis of results. To evaluate the performance of the shield, it is necessary to obtain the field strength readings at the observation points for each of the five cases listed above. The frequency responses of these five simulations are shown in Figures 7.25 through 7.28. For each graph, the unshielded case is plotted to provide a reference.

To derive full benefit from the model, it is necessary to interpret the output data. First, it can be seen that without the shield in place (Figure

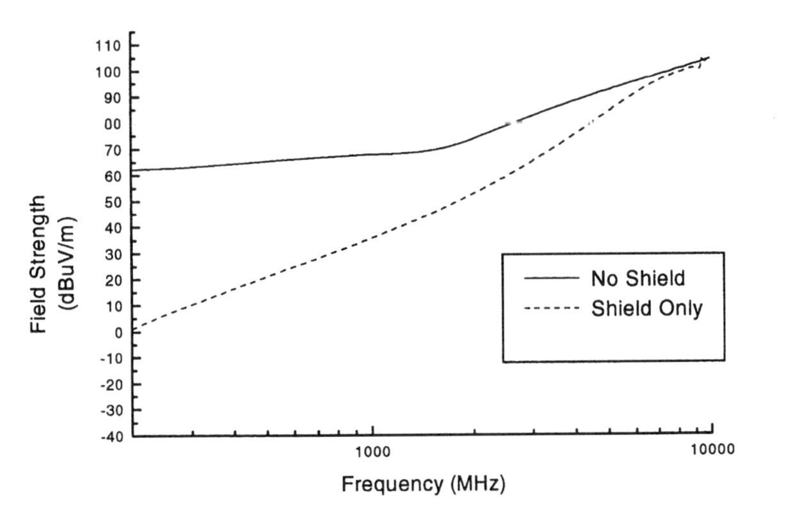

Figure 7.25 Field Strength With and Without the Shield in Place

7.25) the field strength does not drop at a fixed rate with decreasing frequency, but rather there is a change in the slope at around 1.5 GHz. This is a result of being close to the source and is an indication that near-field conditions apply. The second (lower) curve shown in Figure 7.25, shows the response with the shield in place, and the degree of shielding obtained can be seen as the difference between the plots.

In the presence of either the internal or external conductors (Figures 7.26 and 7.27), there are clear peaks at those frequencies where the shielding is greatly impaired. This emphasizes that the effects of such conductors must be considered carefully. In this example the extra conductors are both small and have simple geometries; in a more realistic problem, such conductors could be large and have complex geometries. Without modeling, there would be little possibility of predicting the shield behavior in the presence of such conductors.

The final plot (Figure 7.28) shows the frequency response with both internal and external conductors in place. The results are not what one might initially expect. While the emission peak of the internal conductor is still evident, the peak due to the external conductor is, in this case, greatly attenuated. The interactions between the shield and both conductors cannot be deduced from the earlier simulations, and a full model with all conductors in place is needed.

By building up a model in this fashion, starting with the basic elements and adding more detail piece by piece, users can acquire a

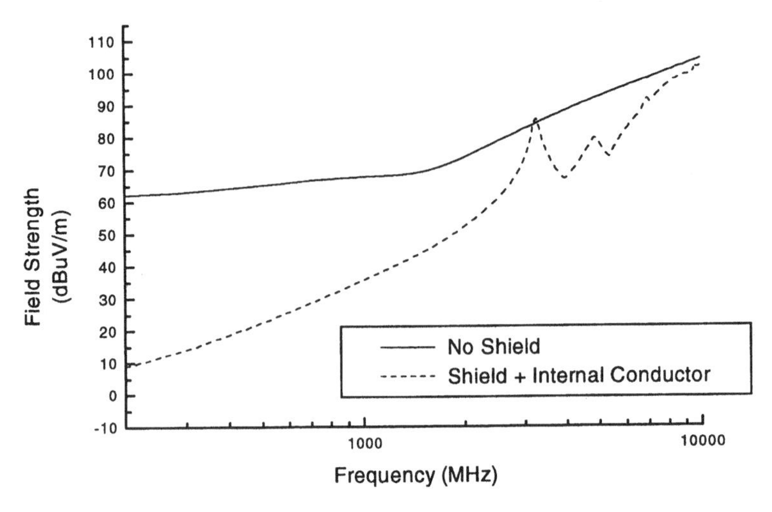

Figure 7.26 Field Strength With Internal Conductor and With No Shield

good knowledge base of what elements are key to increasing and decreasing the effectiveness of the EMI/EMC shield. Through the use of these multiple models, unique and specific EMI/EMC controls will be discovered that can result in significant cost savings for complex designs.

The model created of the small circuit board with a single VLSI

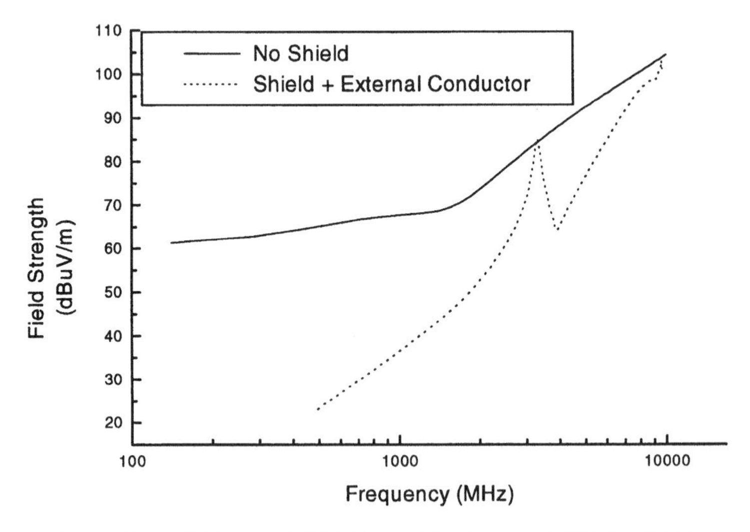

Figure 7.27 Field Strength With External Conductor and With No Shield

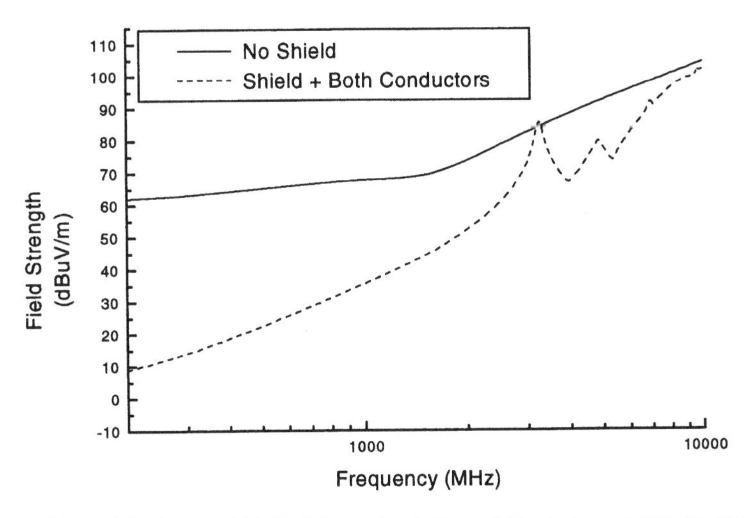

Figure 7.28 Field Strength With Both Internal and External Conductor and With No Shield

device and its heat sink is an excellent example of a specific source model that can be used in conjunction with a shield model as described above. As good airflow is required near devices that dissipate heat, it is not uncommon to find primary sources of EMI located close to the primary sources of enclosure leakage. It has been shown how the shielding performance of a set of apertures is degraded by the presence of a simple single conductor; the presence of a second conductor made the performance quite different. For the case of a more complex structure, such as the circuit board, there is no possible way to predict the overall emissions accurately without considering the coupling effects between the circuit board and shield. This is one area in which the payback for using EMI/EMC modeling is highest as, without modeling, there is little choice other than to build experimental models to find out what might happen. Unfortunately it is very common for this information to be needed before any hardware is available during the very early design stages of a new system.

7.4 Summary

This chapter presents the steps required to create practical EMI/EMC models for different computational techniques. In addition, examples

Table 7.3 Steps for EMI/EMC Model Creation

FDTD	FEM	MoM
Define cell size	Enter problem geometry	Enter problem geometry
Select boundary conditions	Add sources	Define frequency range
Size computational domain	Add observation points	Add sources
Enter problem geometry	Select boundary conditions	Add observation points
Add sources	Define meshing	Define meshing
Add observation points		

of simple yet practical problems were presented to illustrate the use of modeling and to highlight some of the most critical areas.

Every modeling task has its own priorities and variations and the creation of EMI/EMC models is by no means a simple process. However, as a starting point, a guide to the steps needed to prepare for modeling using the three main techniques (FDTD, FEM, and MoM) is given in Table 7.3.

Creating complex models can be done but, to ensure their validity, it is often easiest to begin with smaller specific tasks. These can then be added together as confidence in the independent parts develops. The full complex model that can fully represent the problem at hand can be created by combining these smaller submodels.

8

Special Topics in EMI/EMC Modeling

8.0 Introduction

The uses for modeling and simulation for fundamental EMC issues have been covered extensively in this book. More complex problems can also be readily addressed through modeling. Of necessity, methods used to solve these problems are also more complex. This chapter is intended to serve as an introduction to the modeling of complex problems with the goal of showing different methods of approaching these problems. Creating appropriate models provides many opportunities for the user to express their creativity and ingenuity.

Multistage modeling is useful when the specific situation to be modeled requires different modeling approaches in order to provide a complete definition of the physical situation. Multistage models may use the same modeling technique or different techniques for various parts of the problem. The use of different techniques is known as hybrid modeling. Multistage modeling permits the simulation for each stage to be optimized, thus increasing the overall accuracy and usefulness of the final result. When multiple stage models are used, it is vital that they be separable. That is, each stage must be able to be modeled without effects from the other stages. For example, a physical feature in a stage 2 model must not affect a stage 1 model, or errors in the overall model will occur.

Filters are often used to contain the EMI/EMC energy that is either conducted or electromagnetically coupled onto connector pins. The design of these filters requires knowledge of the expected input and output (I/O) impedance at the frequencies where the filter is to be working. Since this impedance is sometimes the radiation impedance of the physical structure of the Equipment Under Test (EUT) and cables, the value of impedance to use is not readily apparent. Modeling can be used to determine the correct impedance across the frequency range of interest, allowing the filter to be optimized for the application where it will be used.

Although determining the radiated fields from an EUT is often the final goal, other information can be very useful to the EMI/EMC engineer that is available as intermediate results from modeling efforts. These intermediate results can show the radio frequency (RF) current distribution across a PCB reference plane, where the greatest RF current density exists on a shielded enclosure of an EUT, provide insight to the field propagation through slots and other apertures, and so forth. These intermediate results can be extremely useful to help the engineer understand the cause of the problem, and determine the 'fix' for the problem.

EMI/EMC test sites exist in a variety of implementations. Although the preferred commercial EMC test site is the Open Area Test Site (OATS), semi-anechoic shielded rooms, GTEM cells, and shielded rooms are all used for various applications. Each site has its own particular measurement environment, and potential errors. Construction of these sites is expensive, and errors in the site design can be extremely costly. Modeling can be used to determine the effect of design trade-offs on the particular site before the construction is begun.

Although previous work has been done to model various antennas, this is an area that can not be ignored by the EMI/EMC engineer wishing to model a product. The real-world measurement will include the effects of the antenna. For example, the antenna will be nonisotropic (although most models will allow an isotropic receive antenna), and will have a finite bandwidth (although not present in most models). So modeling can be used to better understand the antenna's performance, and how it will affect the final measurement.

8.1 Multistage Modeling

The use of multiple stages to model an EMI problem allows the engineer to solve much more difficult problems than can be otherwise solved using conventional single modeling techniques. As stated in earlier chapters, each modeling technique has certain strengths where it performs well and provides accurate results, but each technique also has weaknesses. It is important to select the individual modeling technique to take advantage of the strengths of the modeling technique and avoid its weaknesses.

Not all multistage models require different modeling techniques. Multiple stages can allow the engineer to use a very fine resolution around an area with small, but important, physical features, and then a second stage can use these results to find the fields using a coarser grid at a greater distance. Regardless, the one important factor is the ability to separate the modeling stages. There must be no "feedback" from the second stage into the first in the real situation, since the first stage will not include the second stage's effects. The individual stages must be electromagnetically separable.

8.1.1 Multistage Modeling for Practical EMI/EMC Problems and Test Environments

Although there is a large number of different types of products that must meet EMI/EMC regulations, most fall into the general class of products with shielded enclosures containing apertures and having long wires attached to the enclosure.[1] Plastic enclosures are often shielded either by a metal internal coating or by metal fragments imbedded in the plastic during the modeling process. Computer products, consumer electronics products, and communications devices all fit this category.

The source of the radiated emissions is usually a high speed (fast rise time) clock or data signal on the printed circuit board within the shielded enclosure. The source creates a complex electric and magnetic field structure within the enclosure. Some of this energy "leaks" out through the apertures (e.g., air vents, slots between option cards, shielded enclosure seams) and creates RF currents on the outside of the shielded enclosure. These currents are then distributed over the entire outside structure, including the wires and cables, and radiate into the outside environment. The fields are then measured 3 or 10 m away in the presence of a ground reference plane.

The products under test typically have long wires attached to different connectors, for example the power cords, modem lines, and printer cables, which will greatly affect the radiated emissions from the product. RF currents that have leaked out from an aperture and are on the outside of the metal shield will couple onto the wires and cables. The wires will greatly increase the effective aperture of the "antenna," that is, the equipment under test or EUT, since the overall size of the EUT with wires is typically increased significantly by the presence of the wires. Experience has shown that products without attached cables and wires can more readily achieve EMI/EMC regulatory compliance. The test standards require that every type of port/interface must be connected to the correct cable or wire, and the cables and wires must be positioned to ensure maximum radiated emissions. Furthermore, the EUT must include all equipment typically included in system configurations, such as computer, printer, keyboard, mouse, and monitor. Figure 8.1 shows

[1]Note: the attached wires are assumed to be connected to the enclosure shield, and are not in the aperture under consideration. Although wiring connectors can be modeled as an aperture with a wire through it, experience has shown that conducted emissions along the wire/connector are a more important concern in those cases. This discussion focuses on radiation through apertures (without wires). Since most exiting wires are intentionally capacitively decoupled to the shielded enclosure for EMI control reasons, they can be modeled as being physically connected to the shielded enclosure. A later section will discuss conducted energy onto a connector (wire through an aperture).

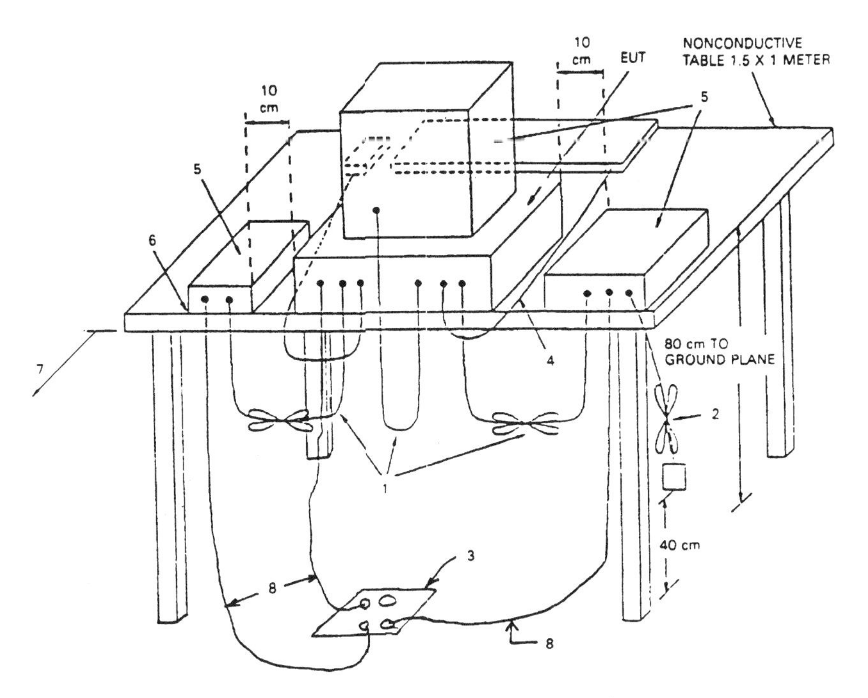

Figure 8.1 Real-World EMI Emissions Test Environment

a typical example of such a computer product with cables. This is a difficult modeling task. Including the test environment makes this modeling task even more complex because the interaction between the ground reference plane and the EUT must be taken into account.

Using contemporary computational technology and techniques, there is no practical way to model the entire problem described above with a single model. Earlier work has been performed that successfully models certain aspects of the overall problem, e.g. radiation from printed circuit boards (PCB) with a microstrip near a reference plane edge [1, 2], PCB via [2], decoupling capacitor placement [5], or shielding through apertures [4–8]; but, because these efforts have addressed only specific facets of the overall problem, they were not adequate to predict compliance with regulatory standards. For example, emissions from an unshielded PCB with a microstrip were modeled [1, 2]. No attempt to include a shielded enclosure with apertures was made. Emissions through apertures in an infinite metal sheet were modeled [4–8], but no attempt was made in these previous studies to include a PCB as the source, nor to include the required measurement environment. These studies were useful to help understand specific phenomena, but did not include all the parts of the overall problem to allow for comparison to the regulatory limits.

8.1.1.1 Emissions Through Apertures

The focus of this section is to demonstrate a hybrid technique to model much more of the overall problem than is possible with only a single modeling stage. The strengths of two different modeling techniques are implemented in this hybrid technique and allow a source, a shielded enclosure with apertures, and the required measurement environment to all be included. Thus, the results of the overall problem can be compared to the regulatory limits for pass/fail analysis. Other internal features, such as partial shielding walls and extra cables, can be included as required. This multistage model uses the Finite-Difference Time-Domain (FDTD) method to model the source and the inside of the shielded enclosure, including the effects of the apertures. The Method of Moments (MoM) approach is used to model the outside of the shielded enclosure, including attached wires, and the test environment.

Modeling the source within the enclosure is a difficult problem. Due to memory and processor speed limitations in contemporary computers, approximations must be made when modeling components within the enclosure. The high-speed traces, bus traces, or CPU heat sinks are typically considered the chief EMI sources and are modeled as relatively simple wire sources, without all the detail of the true printed circuit board. Depending upon the design, separate models might be used to determine the primary emissions sources on a printed circuit board, and then the circuit board is reduced to a simple model with only that source. In order to demonstrate this technique, a PCB edge is assumed to be in close proximity to an enclosure aperture. It has been shown [3] that RF currents along an edge of a PCB reference plane due to microstrip or stripline currents are equivalent to a thin wire (replacing the PCB edge) with a dipole-like current distribution. This is assumed to be the worst case EMI source, but other sources can also be used. The hybrid technique is not dependent on a given source type. The power of this technique allows sources to be modeled as needed. In this example, a simple wire with a current on it is used as the initial source. Simplifying the complex PCB source into a wire with a current has been shown through practical use to provide good first-order results [1, 9, 10].

Once the source model has been developed, the next problem is to model the amount of energy leaking through the apertures in the shielded enclosure due to the fields within the enclosure. For example, the rear panel of a computer enclosure typically has openings which appear as large electromagnetic apertures, often 10 cm long with wires attached nearby.

In order to achieve the required solution accuracy, it is not sufficient to simply model the aperture alone, or even an aperture with a wire in close proximity, but rather the system and complete test environment must be modeled together. Although Figure 8.1 shows a number of different interconnected shielded enclosures, single enclosure products are also tested alone. For the purposes of this example, the problem in Figure 8.1 is restricted to a single shielded box, with a single aperture placed over a ground plane, with the measuring antenna 10 m away. A hybrid model is created to demonstrate that this modeling approach can be used to effectively model this complete problem.

FDTD is used to model the enclosure and all it contains to find the resultant fields in the aperture. The inside of the enclosure can be as complex as necessary and, in addition to the source, can include partial internal shielding walls, compartments, and whatever other features are important. FDTD was selected because it is well suited to modeling this small yet highly detailed part of the problem.

MoM is used to model the outside of the enclosure. It is straightforward to adapt this model to a more complex case by adding external wires or even other enclosures. The MoM was selected for this stage of the problem because it is a surface technique and so only meshes the outside of the structure. Additionally it is not overly expensive in terms of computational resources to add multiple far-field observation points.

The link between the inside FDTD model and the outside MoM models is the fields within the aperture. The field values in the aperture, which are a result of the excitation within the enclosure, are found using FDTD. These values are then used as the source for the MoM model. Since the fields in the aperture are much more dependent on the internal fields than the external fields, the condition that the two models can be electromagnetically separated is met and the multistage model can be considered valid.[2]

8.1.1.2 Hybrid Technique Example

The first step in using the hybrid modeling technique as described here is to create the stage 1 (FDTD) model of the enclosure, aperture(s), the internal source and whatever internal structure is considered important. In the case of an empty shielded enclosure (100-mm cube) with a 10×2-mm aperture, the FDTD cells must be small enough to describe the aperture correctly. For this example, a FDTD cell size of .5 mm is selected. This size is also small enough to provide at least 10 cells per wavelength up to the highest frequency of interest, as per FDTD requirements.

The source is selected to be a simple current on a wire. As described earlier, this is representative of a PCB ground reference plane edge. The wire is oriented perpendicular to the aperture to ensure maximum possible emissions coupled through the aperture.

Figure 8.2 shows a diagram of the FDTD model. The aperture is placed on the top face of the enclosure for convenience, but could be on any side desired. The top part of the enclosure is extended beyond the enclosure walls to restrict any external resonances from affecting the fields in the aperture. The internal structure of the enclosure is maintained to allow any internal resonances to occur. The FDTD simulation must be run long enough to ensure that the ringing of the fields due to the resonance effects have died down.[3] Both the electric and magnetic field at the center of the aperture is saved as the output from this stage one model.

While a detailed understanding of computation techniques is not required for proficient modeling, a sound knowledge of electromagnetic theory as it applies to the problems being modeled, is important. It is always necessary to evaluate the results of a new model to determine

---

[2]If there is doubt regarding the isolation between any two models, it is prudent to make one or more additional FDTD models with different external features and see if the fields in the aperture vary significantly.

[3]The length of time needed to allow the resonances to die down depends on the Q-factor of the enclosure. The use of nonperfect conductivities for the enclosure walls will greatly reduce the Q-factor of the enclosure and provide more realistic results.

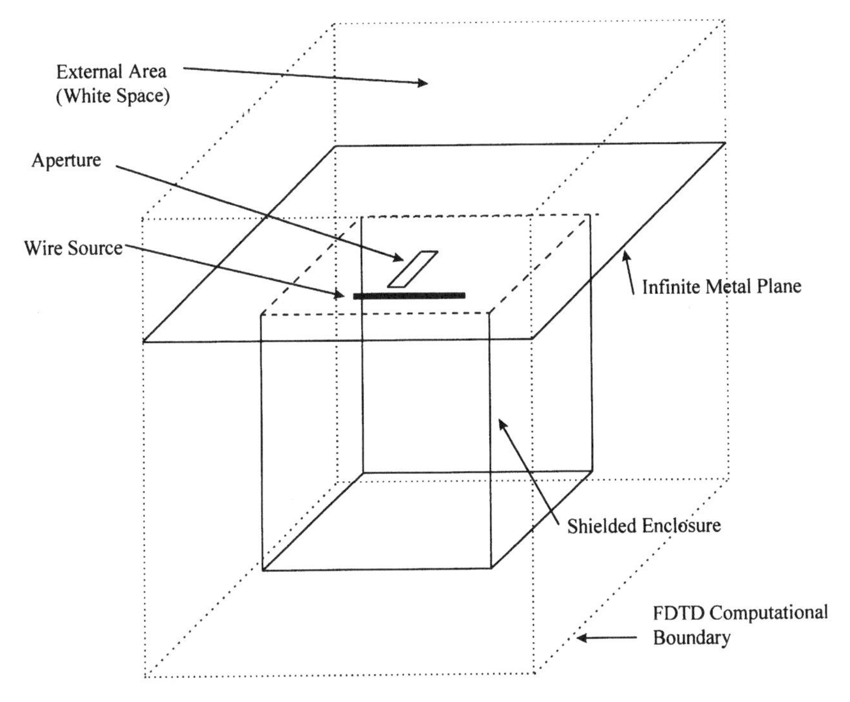

Figure 8.2 FDTD Example Model of Shielded Enclosure and Aperture

whether the behavior is as expected. This can seldom be done for complex structures at frequencies where the structure is electrically large but is often possible for lower frequencies. With FDTD models, absorbing boundary conditions are used to truncate the computational domain. These ABCs become less effective at low frequencies since they are electrically much closer to the source of the fields, and can therefore be a source of error. The impedance of the aperture will be used for an evaluation of the model's low-frequency validity.

The time-domain electric and magnetic field results in the center of the aperture are converted to the frequency domain using a Fast Fourier Transform (FFT). The frequency domain impedance (E/H) is shown in Figure 8.3 and should be examined to determine if errors occurred due to the ABC's close proximity to the aperture at low frequencies which are of interest. In this case, it is observed that at low frequencies the impedance is not dropping steadily as would be expected from antenna theory. This indicates that there is an error which must be addressed.

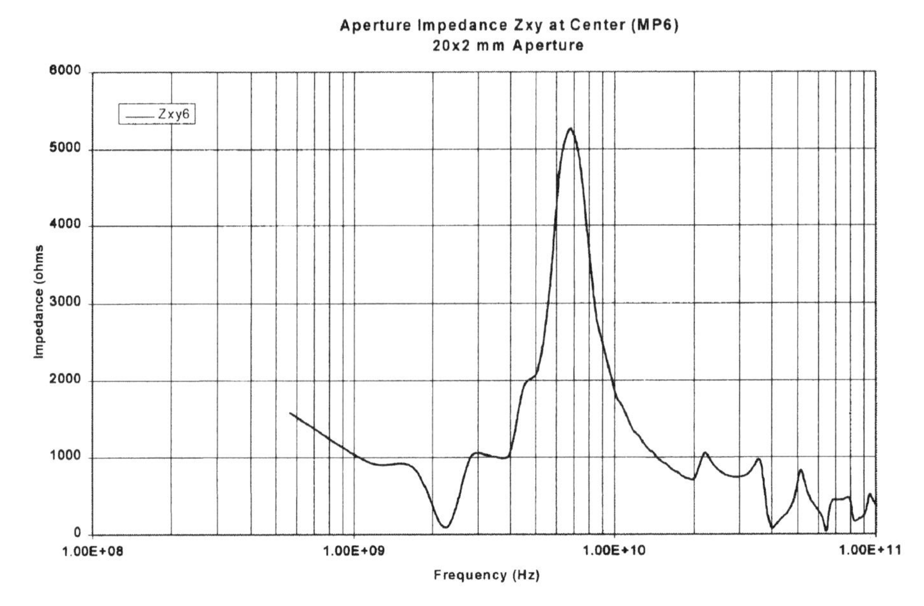

Figure 8.3 Original Impedance for 20×2-mm Aperture

8.1.1.3 Correcting the Low frequency Stage One FDTD Electric Fields

Since the aperture impedance at low frequencies found using the stage one FDTD model appears to be incorrect due to errors introduced by the too-close proximity of the ABCs (and other sources), it is necessary to find a means to correct the field values in the aperture. Because the slot is electrically short, a method using the behavior of a Herzian dipole is used. This correction can be omitted when the FDTD model results are in error at frequencies outside the range of interest to the specific project. However, if the low frequencies are of interest, as in most EMI/EMC activities, the field levels must be corrected or errors will occur in the final results.

Close examination of the aperture impedance curves showed that at frequencies above where the ABC's error was introduced, the impedance varies according to expectation. Experiments with larger FDTD computational domains, and thus greater distances between the ABC and the aperture, shows this region to extend to lower frequencies and so confirms the source of the error.

In the case of a long, thin aperture, Babinet's principle allows the aperture to be represented by an equivalent sized linear antenna. Since aperture impedance is incorrect at frequencies where the wavelength is long compared to the aperture size, the equivalent dipole antenna will be electrically short. The radiation impedance for a short, Herzian dipole is given in equation (8.1). The radiation impedance from the Herzian dipole can be normalized and used to predict the correct impedance in the aperture at low frequencies.

$$Z_{Pred} = 80\pi^2 \left(\frac{l}{\lambda}\right)^2 \tag{8.1}$$

where l is the length of the Herzian dipole.

The impedance at low frequencies (below 4 GHz) is found using the normalized Herzian dipole radiation impedance. The corrected and original impedances are shown in Figure 8.4.

The computational domain size can be increased to observe the effect on the impedance at lower frequencies and ensure equation (8.1)

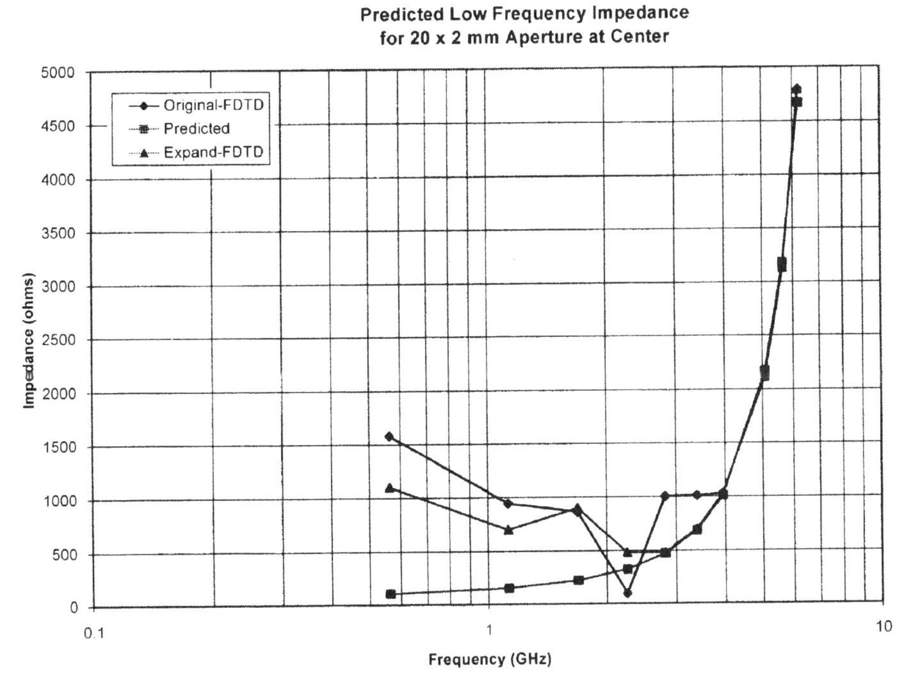

Figure 8.4 Original FDTD and Corrected Aperture Impedance for 20×2-mm Aperture

correctly predicted this impedance. Figure 8.4 shows the impedance in the center of the aperture when the FDTD computational domain is expanded to double the original size. Although this expanded computational domain is impractical, this FDTD model maintains the previous good agreement and extends the good agreement to lower frequencies (before the ABC's again corrupt the fields, and hence the impedance).

Although Herzian dipole impedance method has been used to correct the impedance at low frequencies, the real goal is to find the correct electric field levels in the aperture for use in the second stage of the hybrid technique. The electric field levels are directly proportional to the impedance. Since the impedance at low frequencies is found using the expected Herzian dipole impedance, the corrected values of the electric field at low frequencies can be found using the same curve normalized to the electric field values.

Figure 8.5 shows an example of the corrected electric field at low

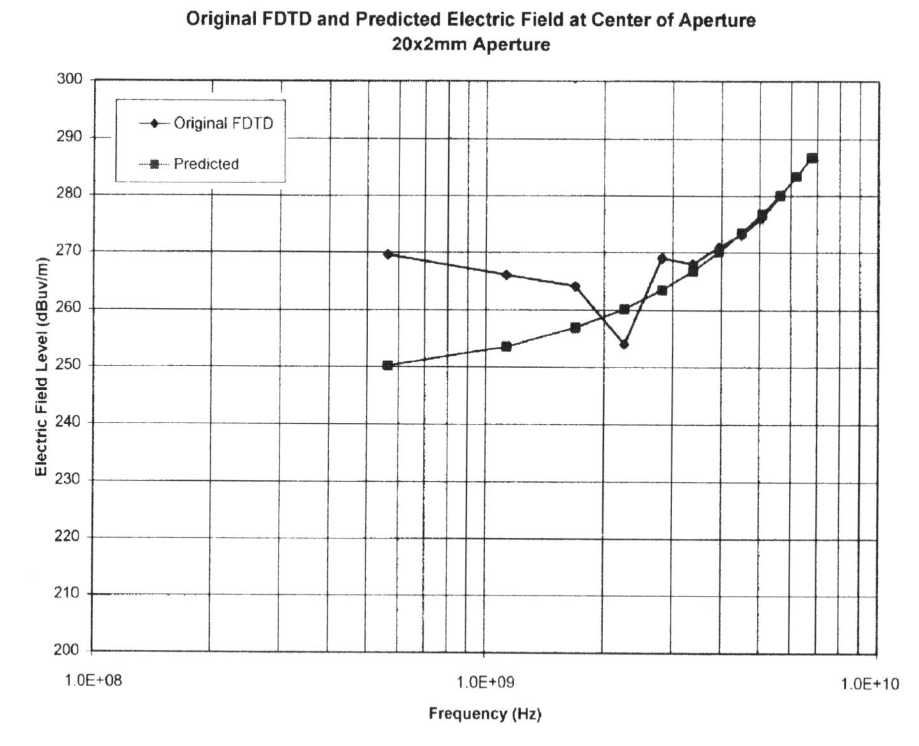

Figure 8.5 Original FDTD and Corrected Electric Field Level in Aperture

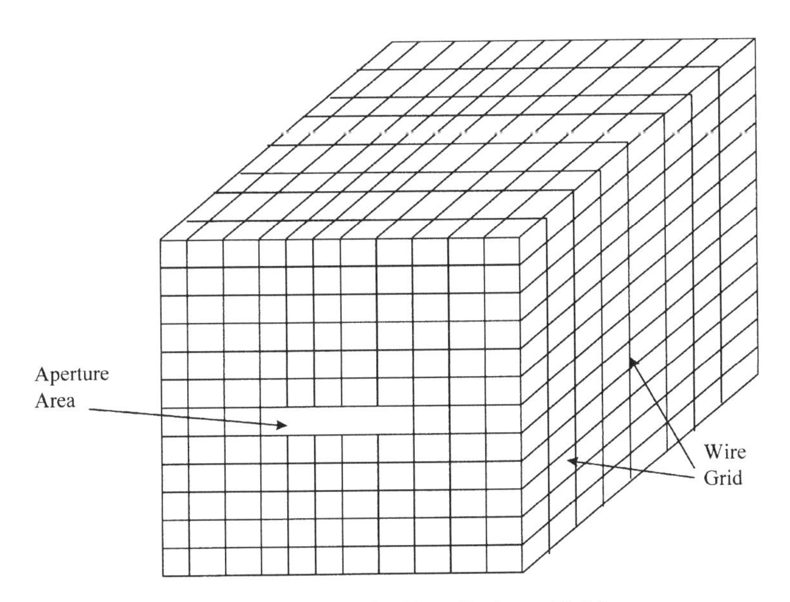

Figure 8.6 MoM Wire Frame Enclosure Model

frequencies using the Herzian dipole technique. The fields from this technique agree with the original FDTD electric field levels at frequencies where the ABC was not contributing an error, and provide the correct electric field strengths at the lower frequencies.

8.1.1.4 External Shielded Enclosure With Aperture Model

The outside surface of enclosure is modeled in MoM for stage 2 of the multistage model using a wire mesh frame with the openings in the wire mesh small compared to the shortest wavelength of interest. The corrected electric field, found from the corrected FDTD results, is then applied across the center of the aperture in the MoM model for each frequency of interest. Figure 8.6 shows a diagram of the wire frame enclosure model. The electric fields can now be found at a ten meter distance from the shielded enclosure.

8.1.1.5 Hybrid Model Comparison Between Free Space and the Real Test Environment

As stated earlier, it is important to model the test environment correctly. The following examples demonstrate the effects of the environment on the final results. As mentioned in Chapter 1, EMI/EMC emissions measurements are required to be made over a ground plane. The receive antenna must be 10 m away, and it must be scanned for maximum receive level over a one to four meter height while rotating the EUT through 360 degrees. The scanning of the antenna height ensures there is no chance of a destructive interference path artificially lowering the measured emissions levels. The rotation of the EUT through the 360 degrees ensures the maximum emissions are received, regardless of any possible directionality of the EUT's radiation pattern.

Figures 8.7 and 8.8 show the model results for the shielded enclosure EUT with and without the ground plane present for both the horizontal and vertical polarizations. As seen in these figures, the presence of the ground plane, and the effect of scanning over the height range, can greatly increase the measured emissions level due to the reflected wave adding in phase to the direct wave.

EMI/EMC emissions test standards also require that all cables be

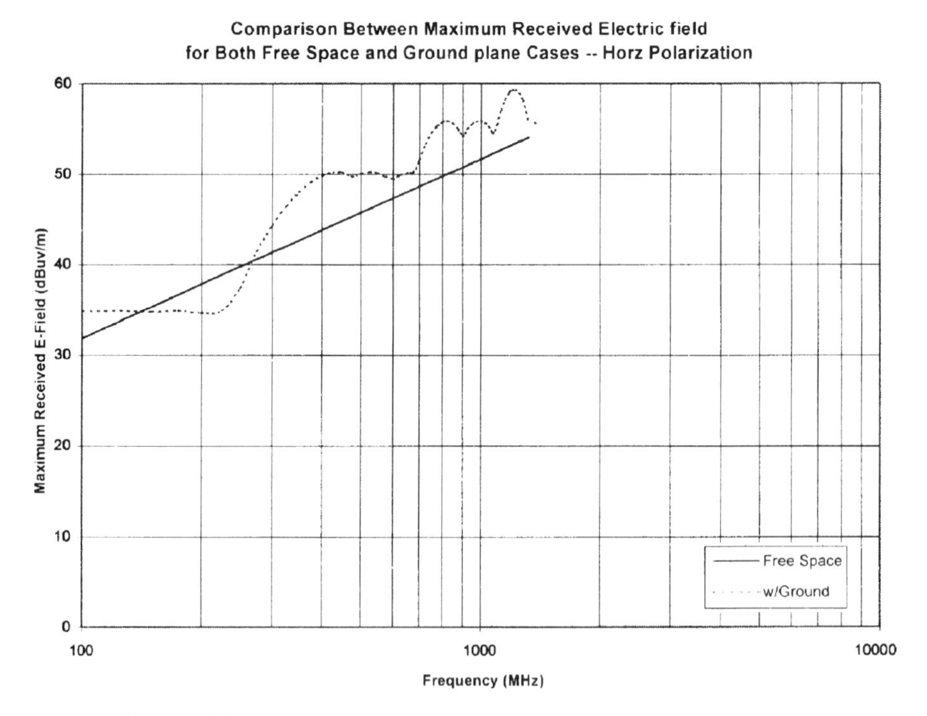

Figure 8.7 Maximized Electric Field Comparison With and Without Ground Plane (Horizontal Polarization)

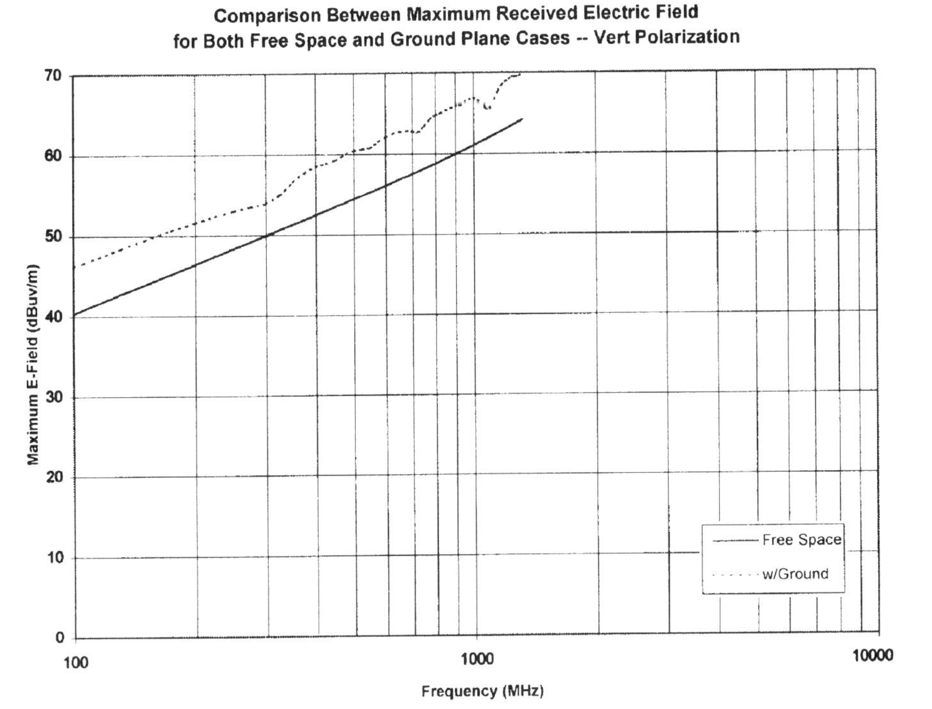

Figure 8.8 Maximized Electric Field Comparison With and Without Ground Plane (Vertical Polarization)

attached to the EUT. This effectively increases the EUT's electrical size, and typically increases the emissions levels significantly at some frequencies. A single cable, one meter long, is now attached to the initial enclosure model, as shown in Figure 8.9. This cable is attached directly to the enclosure shield, as in the case of a cable shield being 'grounded' to the case.

The same electric field is applied across the aperture for this new configuration. The maximum received emissions are greatly increased, as seen in Figure 8.10, due to the addition of the cable. This demonstrates the importance of the hybrid model including all of the test environment features.

8.1.1.6 Summary

This section has shown an example of a hybrid model. This model was expanded from a shielded enclosure with an aperture alone by

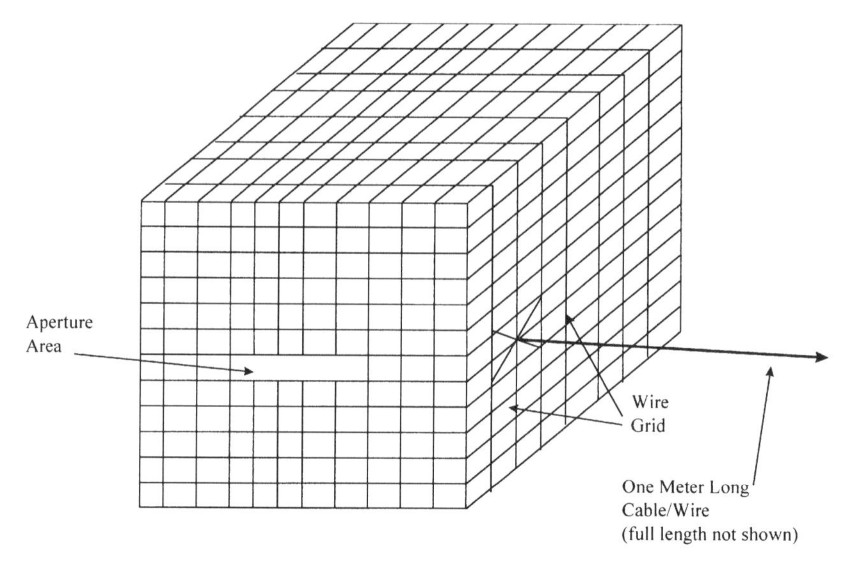

Figure 8.9 MoM Model of Shielded Enclosure with 1-m Cable Attached

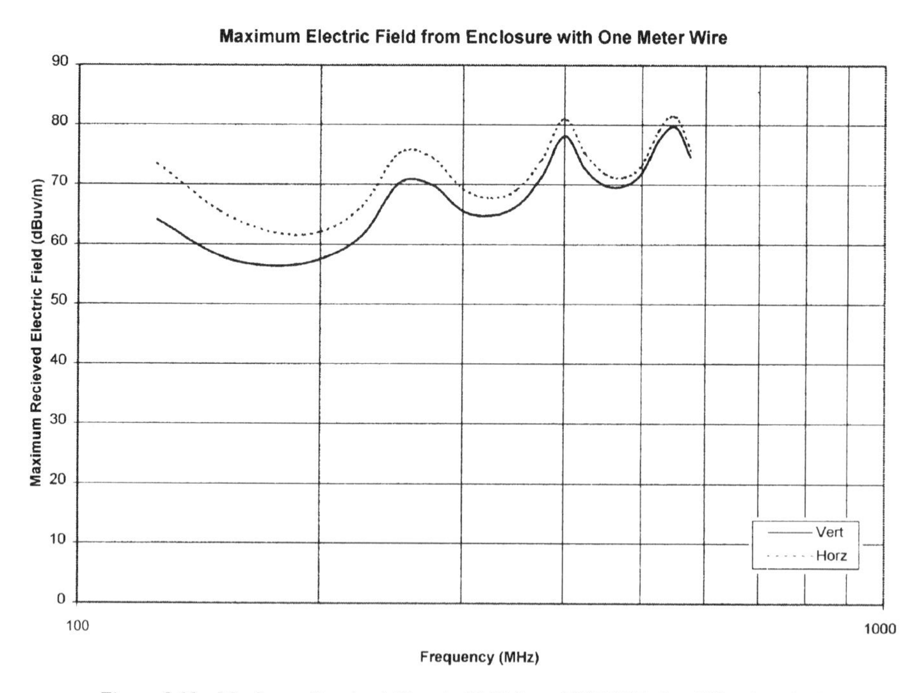

Figure 8.10 Maximum Received Electric Field from EUT With 1-m Wire Attached

adding a long wire. The source is within the enclosure, and the real-world measurement environment is included in the model. An FDTD model is used as a stage 1 to model the source inside the enclosure and to find the fields within the aperture. These fields within the aperture are then corrected at low frequencies using the Herzian dipole impedance technique.

Once the fields within the aperture are known, they are used as the source in the MoM second stage by placing a single voltage source across the center of the aperture. Complicating features, such as a long attached wire, can be easily included, as well as measurement environment features such as the ground reference plane, 10-m measurement distance, rotation, and height scanning. The inclusion of all these features allows the final result to be directly comparable to the measurements that are made at an open area test site.

8.1.2 Emissions Through Apertures With Wires (Connectors)

Another good example of a typical EMI/EMC problem is a printed circuit board in a shielded box, with a long wire attached to a connector mounted in the shielded box and connected to the circuit board. The amount of EMI noise from a source inside the enclosure, for example, some signal traces on the circuit board, which couples to the connector pins, is conducted to the outside of the shielded box, and radiated from the cable, is to be predicted. Since an inside-outside problem is present (leakage through the shielded box's aperture on the connector pins) the FDTD or FEM technique should be selected. However, the long wire attached to the connector would be best modeled using MoM! MoM would not serve well for the inside/outside part of the problem, nor would FDTD serve well for the long wire part of the problem. This is a very difficult problem and extremely cumbersome to solve with only a single model. However, this problem can again be solved using multiple stages, in a fashion similar to the previous aperture problem.

A natural separation point for this problem is at the aperture where the connector wires leave the shielded enclosure. The amount of coupling between the signal traces and the connector pin in the aperture can serve as the first stage model, and the common mode voltage on the connector pin (relative to the shielded box) is the output from the first stage. The first stage can be solved using FDTD for the same reasons

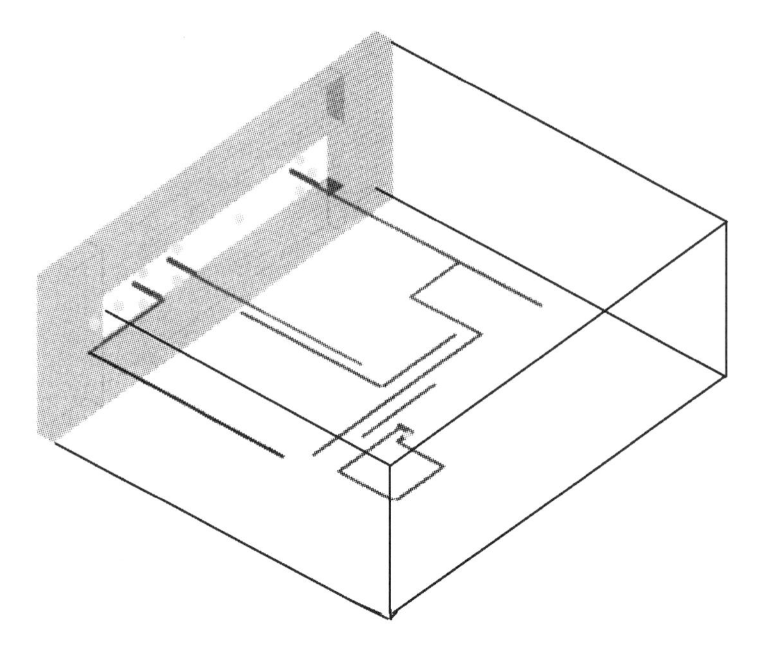

Figure 8.11 PCB Model (Cutaway View) for Inside Problem (Stage 1)

as the previous example. An example is shown in Figure 8.11 of such a model.[4] In this case the FDTD simulation code uses a modified Liao absorbing boundary condition (ABC) and the external long wire was simulated by running a short wire straight into the ABC, simulating an infinitely long wire. This assumption is reasonably good when the wire is assumed to be infinitely long. (See Section 8.1.2.1 for electrically shorter wires.) The output from the first stage is found by using the worst-case (highest levels) electric field between the connector pins and the shield. Knowing the distance between the pins and the shield determines the common-mode voltage on those connector pins. In this case, FDTD monitor points are positioned between the various pins and the shield to find which connector pin has the highest common-mode voltage.

The radiation outside the shielded box is most likely due to the effects of the long wire increasing the effective antenna size. The common mode voltage between the cable and the shielded box is the

[4]Note: the PCB reference plane and some of the enclosure has been removed for clearer viewing of the source traces and the connector area.

source. This model can be solved using MoM and the shielded box either broken into surface patches or wire frame segments. An example using a wire frame model is shown in Figure 8.9. In this case, the long wire runs horizontally for one meter, then drops to the ground plane. The fields can then be found at the desired distance, scanned height and rotation. The effects of a ground plane are typically included in EMI/EMC measurements, and can be included in this MoM simulation.

8.1.2.1 Simulation for Electrically Long External wires

As stated earlier, the effects of a long wire on the outside of the shielded box must not be completely ignored while solving the coupling inside the shielded box. However, since no modeling technique can easily solve this complete problem, it must be broken into individual stages for solution. When the external wire is electrically very long (λ << wire length) it is necessary to somehow represent this in the FDTD model. It is possible to make an infinitely long wire in FDTD by extending one end of it into the FDTD ABC, as described in the previous section. However, this approach will not give good results for cases where the wire can not be assumed to be infinitely long.

A preliminary step is needed to find the correct impedance of the external wire at the connector. In this example, the MoM is used to determine the wire's load or radiation impedance. This impedance can then be used as the 'load' impedance for the FDTD enclosure model. This impedance may be implemented in FDTD by using the Surface Impedance Boundary condition (SIBC) technique, or a number of individual FDTD simulations can be run using the impedance at discrete frequencies.[5]

Many models use a normalized source spectrum (typically a Gaussian pulse for FDTD or a constant voltage at each frequency for MoM); the final result is then de-normalized to include the true input source spectrum. Therefore, the order of the simulations are not important. So, the second stage model can be solved independently of the output of the first stage (again, the necessary condition for the two stages to be independent remains). As you will recall from Chapter 4, the MoM

[5]The FEM may also be used for the stage 1 "inside" model since the wire load impedance is provided at discrete frequencies. However, this may not be sufficient to find all the internal enclosure resonances.

technique finds the currents on every segment or patch in the model. If the source is applied at the segment between the shield of the box and the long cable attached (as in the case of a common mode voltage source), the ratio of the voltage to the current at the source spectrum provides the source impedance. Since both stages are (now) using a normalized source spectrum, each only needs to be solved once; the MoM portion (previously called stage 2) is solved first, and the FDTD portion (previously called stage 1) is solved second.

8.1.2.2 The Final Multistage Model Result

The final result from the multistage models is found most easily using a spreadsheet utility program. First, the result of stage 1 must be de-normalized, to account for the true source spectrum, as shown in Figures 8.12a and 8.12b. The de-normalized output of stage 1 becomes the input spectrum for Stage Two, however, the stage 2 model used a normalized spectrum also. Therefore, the output of stage 2 must be de-normalized to its "true" input spectrum, which is found previously by de-normalizing the output of stage 1. So the final result is the output from stage 2, after both de-normalizing steps.

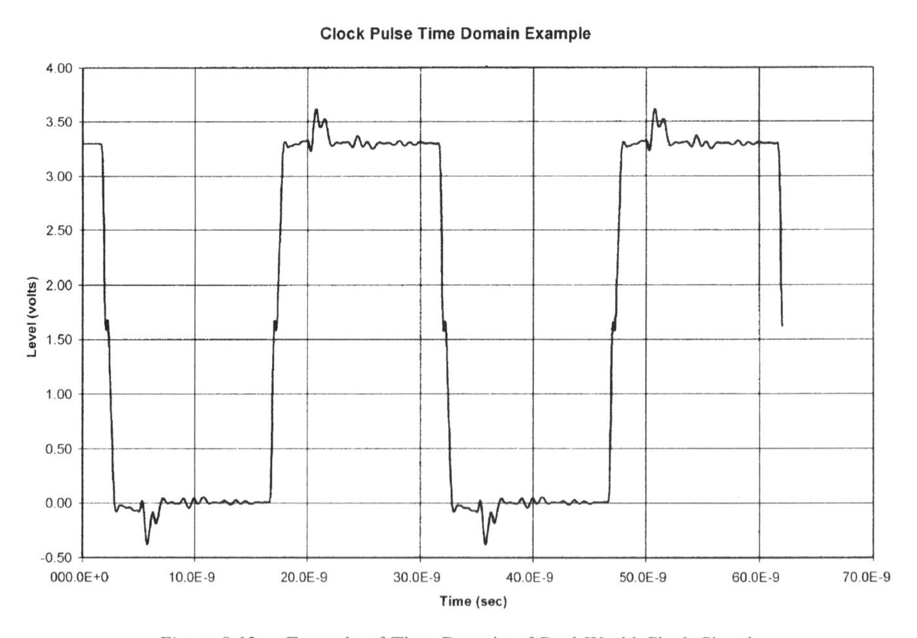

Figure 8.12a Example of Time Domain of Real-World Clock Signal

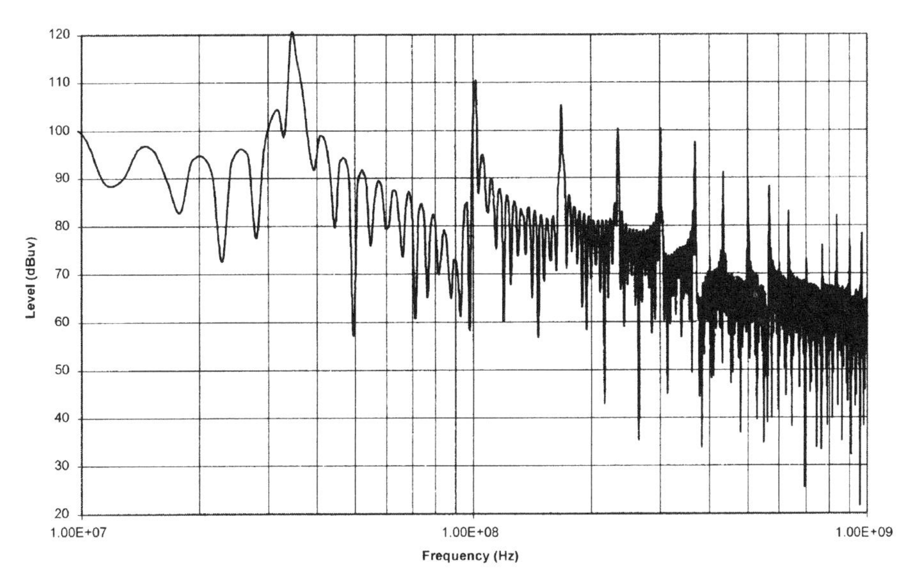

Figure 8.12b Example of Frequency Domain of Real-World Clock Signal

8.2 Designing EMI/EMC Filters

It has been stated a number of times in this book that a cause of EMI/EMC emission problems is the long wires or cables attached to the primary equipment containing common mode currents at frequencies never intended or needed on that cable or wire. This long cable or wire creates a more efficient antenna (whether for emissions or immunity).

Cable/wire shielding is often used to help reduce the exposure of these common mode currents to the outside world, but this is an expensive option, and nonshielded cables are usually preferred and sometimes required. Therefore, a filter must be used to reduce the emissions conducted onto the unshielded cable. Numerous books are available on filter design, and this information will not be repeated here. However, the performance of any filter is dependent on the I/O impedances, and these impedances are typically difficult to know in EMI/EMC applications. Unfortunately, many engineers simply ignore the importance of these impedances' importance, and design filters for 50-Ohm I/O impedances. While 50 Ohms is as good as any other arbitrary

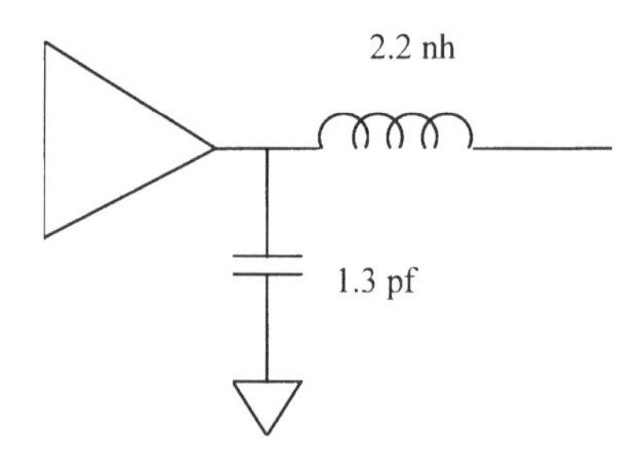

Figure 8.13 Output Impedance Example of Typical IC

number,[6] it does not include the reactive portion of the real-world impedance.

8.2.1 Filter Input Impedance

For the emissions problem, the input impedance the filter is likely to "see" is due to the printed circuit board and components on the board.[7] Often the input impedance for a filter is a simple matter to determine. For example, if the filter is connected to an impedance controlled trace, e.g., microstrip or strip line traces, the filter's input impedance will be that controlled impedance when the trace is electrically long. When the trace is not electrically long, the impedance will be the device driver impedance through the transmission line.

Often the signals connected to the outside world are not sent via impedance controlled traces, but rather by a simple (usually short) PCB trace. The input impedance for the filter is now determined from the driver in the integrated circuit that is used to drive the external wire. An example is shown in Figure 8.13. As can be seen, the impedance of the driver is due to the capacitance and inductance, and the impedance across the frequency range can be found until the parasitic inductance and capacitance become more important than the driver specific impedance.

8.2.2 Filter Output Impedance

As in the input impedance case (indeed as in ALL cases involving a model), the detail required in the model depends upon the highest

[6]Filter design text books will often suggest an output impedance range between 10 and 200 ohms.

[7]Note, for immunity problems, this impedance becomes the filter output or load impedance.

frequency of interest. Typically, good EMI/EMC design practice will dictate the output I/O driver be placed as close to the filter and output connector as possible. So the filter's output can usually be considered to be the connector pins. The output impedance of the filter becomes the impedance from the long wire and the shielded box, which creates an asymmetric, dipole-like antenna. There is no way to directly determine this impedance.

The MoM technique is useful to find the impedance of the shielded box and wire. As you will recall from Chapter 4, the MoM technique finds the currents on every segment or patch in the model before it is able to find the radiated electric fields. If the source is applied at the segment between the shield of the box and the long cable attached (as in the case of a common mode voltage source), then the ratio of the voltage to the current at the source spectrum provides the source impedance. This source impedance is actually the filter's output load impedance. The EMI/EMC filter can now be designed using the correct values of both the input and the output impedance, thus eliminating filter over-design.

As an example, consider the external structure shown in Figure 8.14. This example has a connector through an aperture, with a long wire attached. The external source is a common-mode voltage between the long wire and the shield of the enclosure. The impedance this external

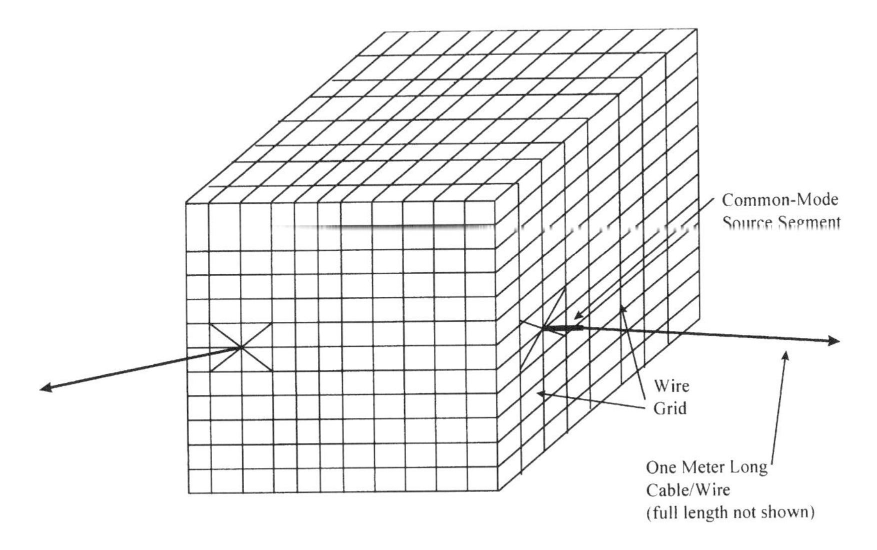

Figure 8.14 External Model for Multistage Models With Connectors

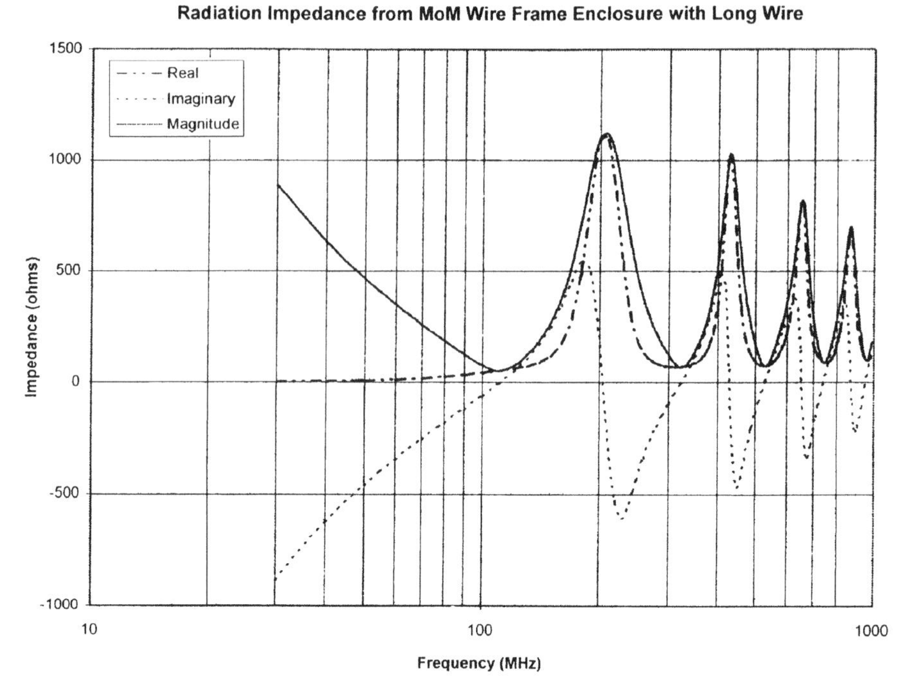

Figure 8.15 Radiation Impedance for Wire Frame Shielded Enclosure With Long Wire Attached

source "sees" is the radiation impedance, and the output impedance to a filter placed just inside the connector. The filter would then reduce the amount of common-mode source voltage.

Figure 8.15 shows an example of the radiation impedance from an MoM model of this structure. Although the impedance varies across the frequency range, typically the problem frequencies can be reduced to a small subset, and the filter's output impedance selected. If the output frequency range is too wide for a simple impedance selection, then the worst case radiation impedance must be used (the lowest impedance) over the entire range.

8.3 Intermediate Model Results

Much focus is given to the radiated fields when discussing EMI/EMC modeling. However, there are some intermediate results from the models that often become very important, and sometimes, more important than the radiated fields results.

8.3.1 RF Current Distribution

The distribution of the RF currents over a ground reference plane or shielded enclosure can be easily imagined by a competent engineer for low frequencies, but is extremely difficult to predict (by anyone) at high frequencies. However, as part of the normal calculation process, the MoM finds the RF currents on every segment and patch in the model, thus providing the RF current distribution.

8.3.1.1 RF Current Distribution on a Printed Circuit Board Ground Reference Plane

Figure 8.16 shows an example of a typical printed circuit board trace over a ground reference plane. In this case, the MoM was selected to model this problem, and the surface patch MoM model is shown in Figure 8.17. A source was placed between the PCB trace and the ground reference plane, and a load placed at the opposite end of the trace. Figure 8.18 shows the amplitude of the RF current at one of the harmonic frequencies of the basic clock frequency (100 MHz). In Figure 8.18, the microstrip trace runs vertically, and is indicated by the area of highest RF current amplitude.

A second example is shown in Figure 8.19, where a split is present in the ground reference plane. This split could be due to a number of different reasons or physical constraints. The resulting RF current density is shown in Figure 8.20. As shown in the figure, the RF current can be seen to travel around the void in the reference plane, to return to the source area.

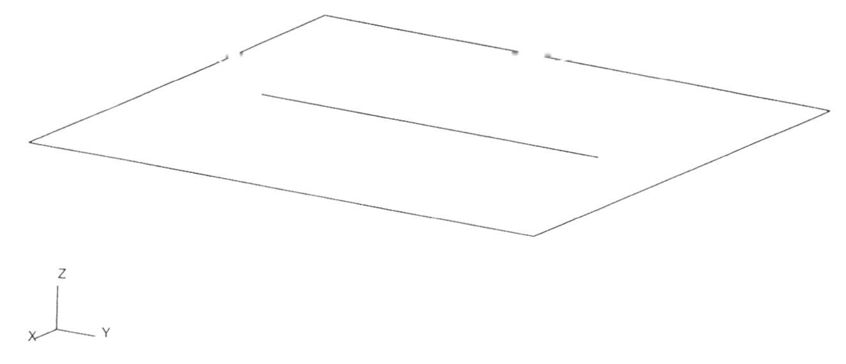

Figure 8.16 Microstrip Trace over Ground Reference Plane

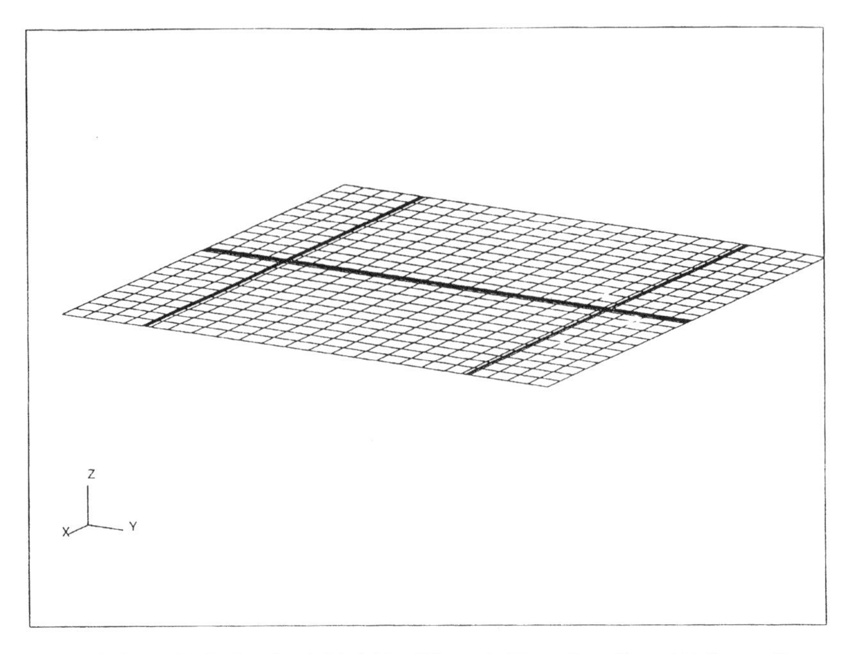

Figure 8.17 MoM Surface Patch Model for Microstrip Trace Over Ground Reference Plane

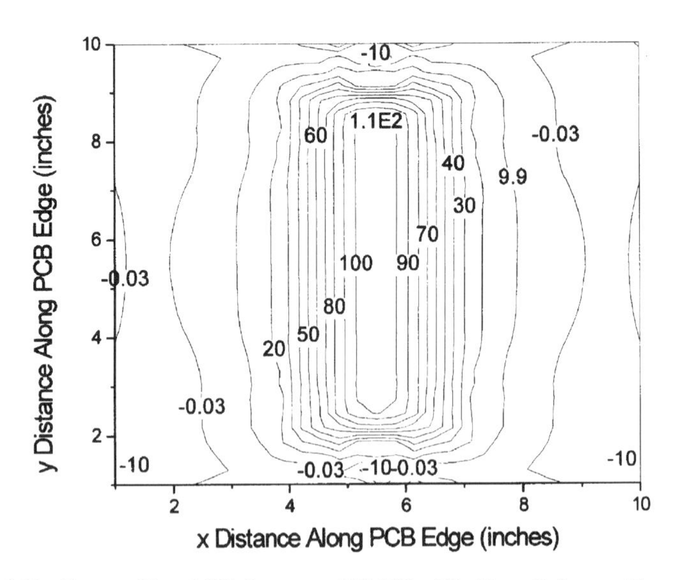

Figure 8.18 Contour Plot of RF Current at 100 MHz (dBuA) on Reference Plane Due to Microstrip

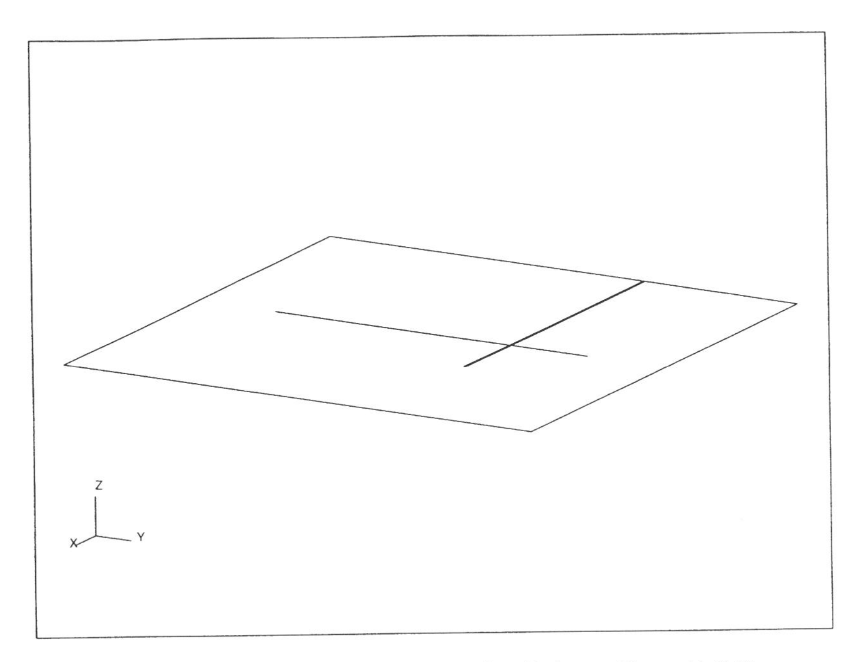

Figure 8.19 Example of Microstrip Trace Over Reference Plane with Split

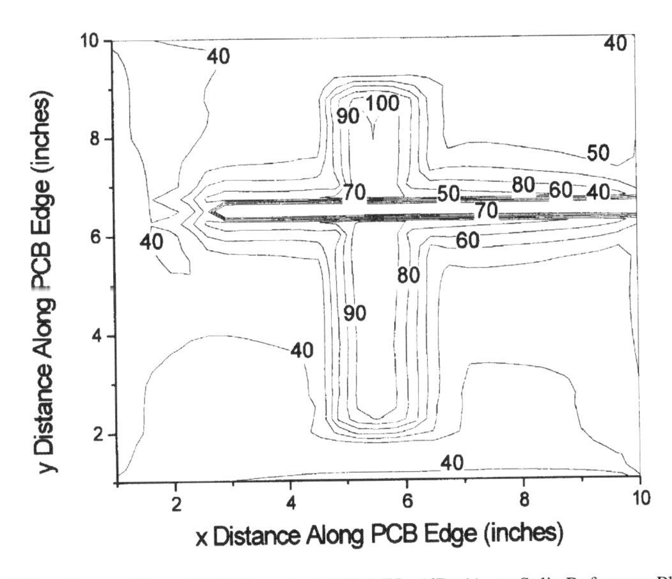

Figure 8.20 Contour Plot of RF Current at 100 MHz (dBuA) on Split Reference Plane Due to Microstrip

These kinds of plots can be helpful in understanding the RF current path, especially at high frequencies, and helping to determine the effect of changes in this current path. Another use for this information could be to determine the likely best location for lumped filter components, such as capacitors and ferrite beads. If the circuit board traces are modeled at frequencies where unintentional electromagnetic coupling occurs, then the areas where this unintentional and undesired currents exist would be easily visible. The areas where these currents exist are obviously the best areas to place such components. Without this information these components may not reduce the EMI/EMC problem, or could even make it worse. (See Section 8.3.2 for more discussion on lumped components.)

8.3.1.2 RF Current Distribution on a Shielded Enclosure

Another application for which being able to view the RF current distribution could be useful is a shielded enclosure with apertures. If the enclosure and its attached cables are electrically short, then the current distribution on the enclosure will be fairly constant. However, once the frequency becomes high enough to cause the enclosure to be no longer electrically short, then the distribution of current on the outside of the enclosure will become complex.

This complex current distribution would be difficult (or impossible) to predict using only intuition. However, it might be important to know the current distribution at critical frequencies so the mechanical design engineer can ensure that any necessary apertures are not placed in such a manner to obstruct the current flow, and create additional emissions. Once areas of high current density are found, they can be avoided for air ventilation slots and other apertures.

8.3.2 Perfect Components

Real-world components are constrained by the "possible." For example, it is not possible to build a capacitor with no inductance or resistance, nor is it possible to build an inductor or ferrite bead with no capacitance. However, such perfect components are possible in the modeling and simulation world.

For example, the MoM technique easily allows lumped components to replace short wire segments. These lumped components can be either pure resistance, capacitance, or inductance. They can also be a combination of the three, or a real and imaginary impedance. However, as a perfect capacitor (zero ohms at the frequencies of interest), the ideal location for a capacitor could be located without concern about the value of the capacitor, or its self-resonant frequency. A perfect ferrite bead (infinite ohms at the frequency of interest) could also be placed at various positions without concern for its capacitance.

This can be a very useful analysis tool. Naturally, once the best location for a perfect component is found, then the perfect capacitor should be replaced with a real value of capacitance, and the series inductance and resistance included. Thus different values of capacitance and even different capacitor types (since different capacitor types/styles will have different series inductance and resistance) can be evaluated before circuit board construction is begun.

8.4 EMI/EMC Test Sites

A number of different EMI/EMC test sites are typically used, depending upon the test standards, the amount of space available for the test site, and the cost of the test site. These test sites include the Open Area Test Site (OATS), the semi-anechoic shielded room, the fully-anechoic shielded room, the mode-stirred chamber, the GTEM cell, the TEM cell, and a (plain) shielded room. A description and example of modeling for some of these applications are presented.

8.4.1 Open Area Test Sites

The construction of Open Area Test Sites (OATS) for commercial EMI/EMC testing for radiated emission requirements is at an all-time high. The demand for test facilities to meet the amount of testing required for computer, consumer, medical, and other products has forced many test laboratories to turn away business due to a lack of test capacity. Semi-anechoic rooms are sometimes used to perform radiated emissions testing, but these sites are very expensive and usually limited to large companies doing internal product testing. Therefore the construction of OATS facilities is rising to try and fulfill the test capacity demand.

The preferred location for an OATS is close to the product development areas, and yet in a wide open, RF quiet environment. These requirements are usually in conflict with each other, since a geographical location with a high concentration of product development activity tends to be located in close proximity to a highly developed area. Land prices are high, and the ability to have large open spaces around the OATS requires often excessive property purchases. As new OATS facilities are planned, the use of existing land is preferred, even if the recommended space and ground plane sizes are impossible.

Once a desired location is identified, and the ground plane size and near-by metal structures are identified, some determination must be made to see if the proposed test site will meet the CISPR site attenuation requirements. Since it is often desirable (or necessary) to violate the ground plane and open space recommendations, it is necessary to find some way to give a high degree of confidence the OATS will be acceptable, without extra expense after construction is completed. This section discusses OATS analysis using numerical modeling techniques. Comparison between various ground plane sizes and shapes, and near-by metal objects, are made to a perfect site.

8.4.1.1 The Perfect OATS

The definition of a perfect OATS would include an infinite ground plane with no other metal structures, e.g., fences, posts, and power lines nearby. Of course, it is impossible to achieve such a perfect OATS in real-life, but a perfect OATS can be simulated using numerical modeling techniques. For this example, both the MoM and the FDTD techniques are used. A small dipole antenna was created such that the length of the antenna was very short compared to the frequencies of interest, 30 MHz to 1 GHz, and the receive location was scanned from a height of one to four meters, at a distance of 10 meters from the transmit antenna.

The case of a perfect (infinite) ground plane was first compared to the case of no ground plane for horizontal polarization. The deviation between perfect and no ground plane for the MoM case is shown in Figure 8.21. The comparison was repeated using FDTD with cell sizes of 100 millimeters, which should give reasonable results to 300 to 500 MHz. The FDTD comparison results are shown in Figure 8.22. In both cases, the distance between the transmit and receive antennas was 10 meters, and the receive antenna was scanned over one to four meters for the maximum electric field amplitude. As can be seen by comparing

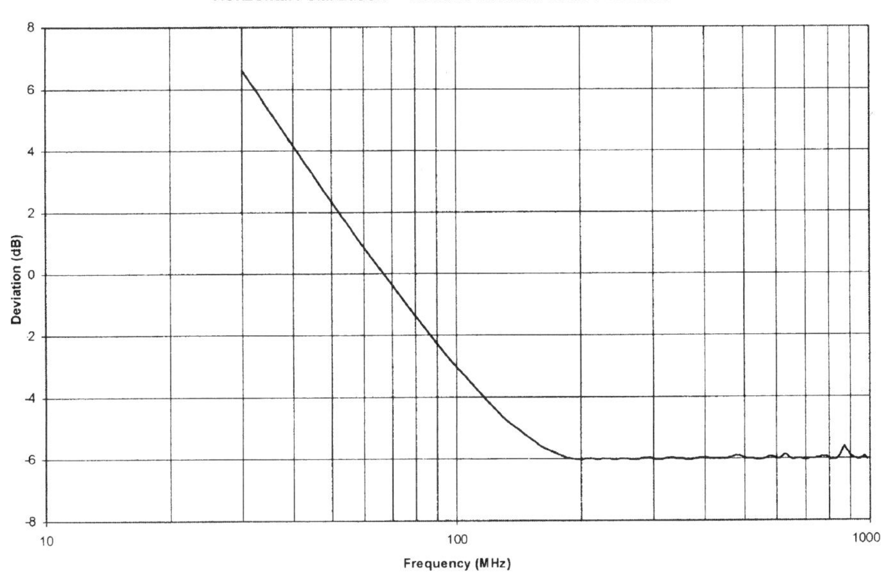

Figure 8.21 MoM Modeled Difference for OATS With and Without Infinite Ground Plane

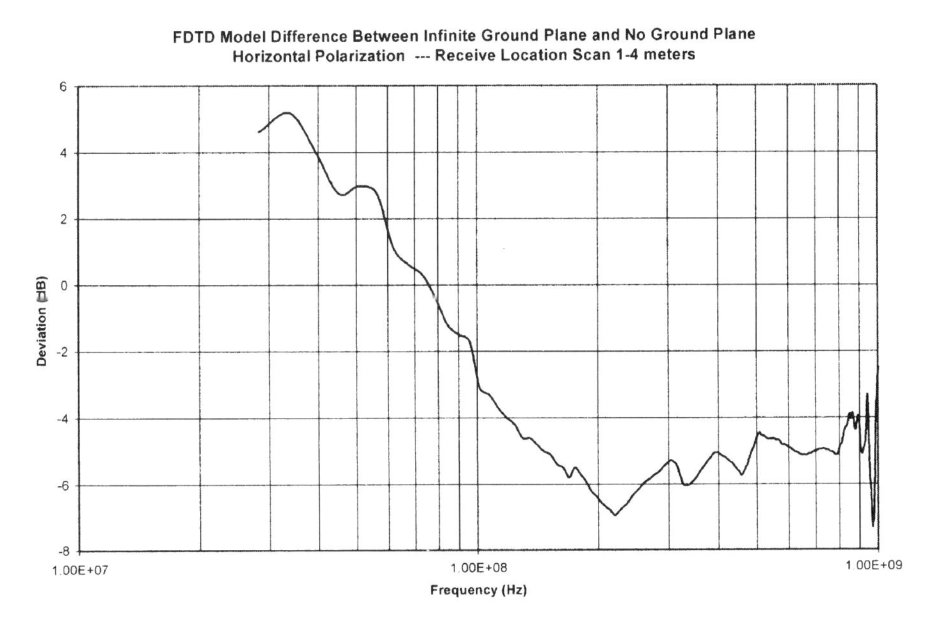

Figure 8.22 FDTD Modeled Difference for OATS With and Without Infinite Ground Plane

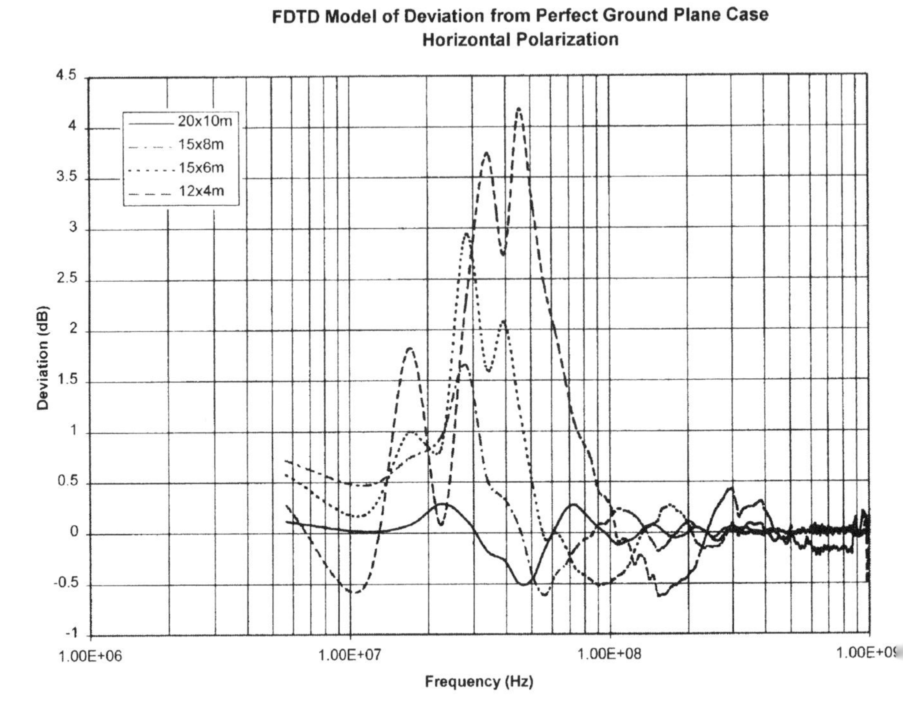

Figure 8.23 FDTD Model With Various Ground Plane Sizes (Horizontal Polarization)

Figures 8.21 and 8.22, the two modeling techniques agreed to within one dB over the frequency range of 30 MHz to about 800 MHz.

The perfect, infinite, ground plane case became the normalized case. That is, all further results are shown as a deviation from this perfect OATS case. Although most of the results are shown for the horizontal polarization, the vertical polarization can be analyzed just as easily, and is used for the near-by metal conductor examples.

8.4.1.2 FDTD Modeling of Ground Plane Size

Although any numerical technique could be used to model a reduction in ground plane size, FDTD was considered the most convenient to use. The deviations from a perfect ground plane for both the horizontal and vertical cases are shown in Figures 8.23 and 8.24 respectively.

8.4.1.3 Nearby Conductors

The presence of metal conductors in the nearby vicinity of an OATS is sometimes a fact of life that is unavoidable. For example, a number

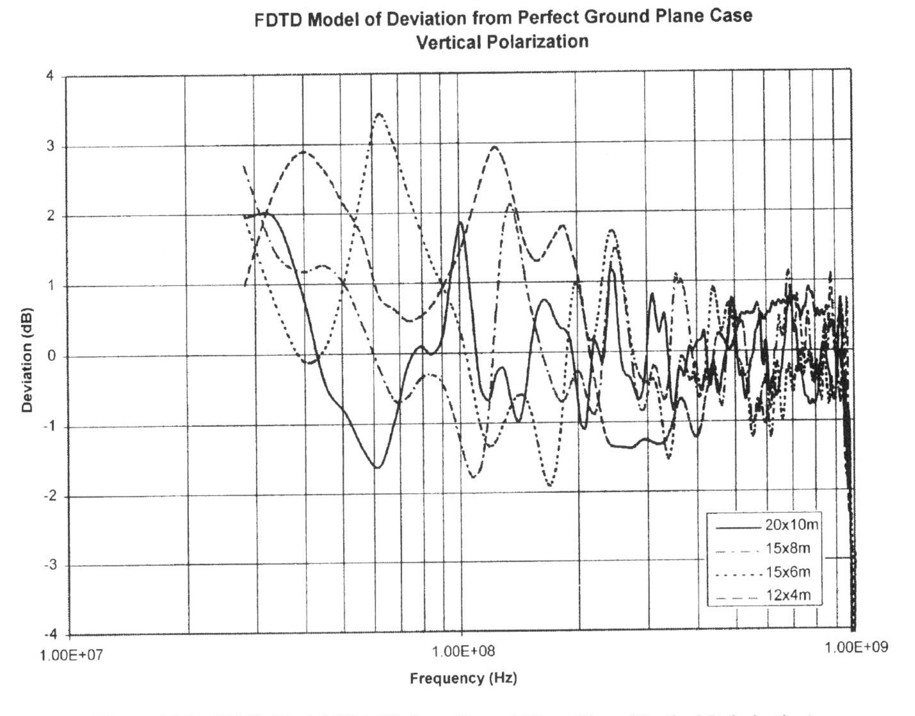

Figure 8.24 FDTD Model With Various Ground Plane Sizes (Vertical Polarization)

of cases could be used to determine the effect of allowing a metal fence either alongside the OATS or behind the receive antenna at various distances. An example is shown of a simulation of a metal light/utility post at two different distances from the EUT side of the OATS ground plane. As shown in Figure 8.25, the effects of the different set-back distances for the posts are clear.

8.4.1.4 Surrounding Walls

Since it is often desirable to put a weather-proof structure around an OATS to allow year-round use, the question often arises of how much conductivity can be allowed in the walls. The FDTD method is particularly useful for materials with limited conductivity or even a relative dielectric constant of other than one. Figure 8.26 shows the results of using FDTD to model an OATS with external walls that were partially conductive. As can be seen from this figure, the walls had a definite effect at low frequencies.

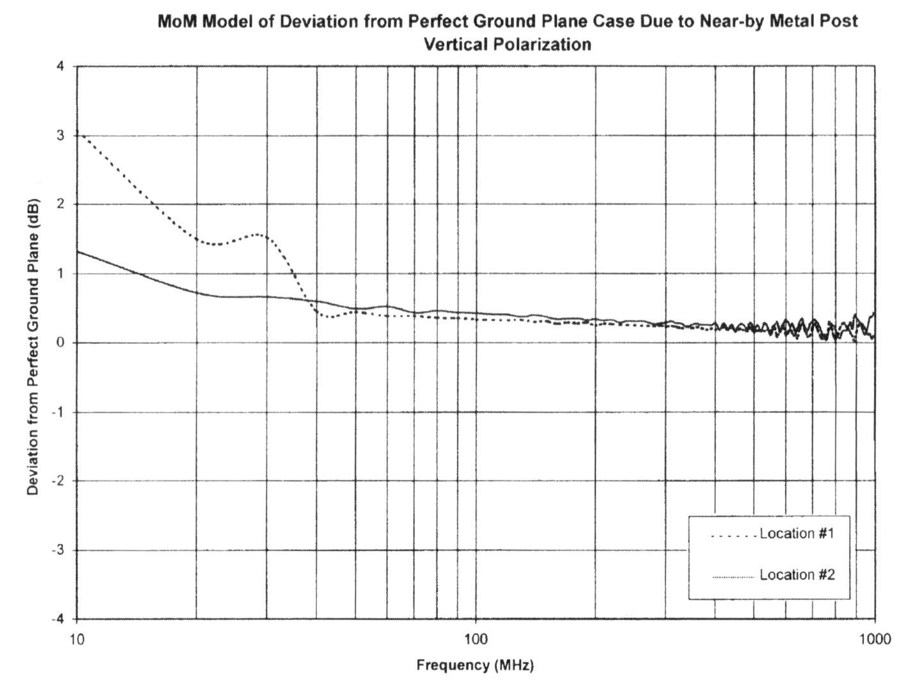

Figure 8.25 MoM Model Deviation Due to Presence of Nearby Metal Posts

8.4.1.5 Summary

The design of OATS facilities sometimes require a deviation from the recommended ground plane sizes or nearby conductor spacing. Since such deviations can result in serious cost penalties if the OATS cannot pass the normalized site attenuation certification tests after construction, normal practice is to avoid any deviation, even though it will result in added cost to the facility during construction.

However, numerical modeling techniques have been shown to be able to help analyze these nonstandard OATS designs and provide engineers with a risk assessment in terms of expected normalized site attenuation error versus cost of design options. The various design parameters that can be analyzed include ground plane size, ground plane shape, and distance to nearby conductors, such as fences and metal poles.

Both FDTD and MoM have been shown to be useful, depending upon the type of problem. MoM requires a wire frame mesh (or a

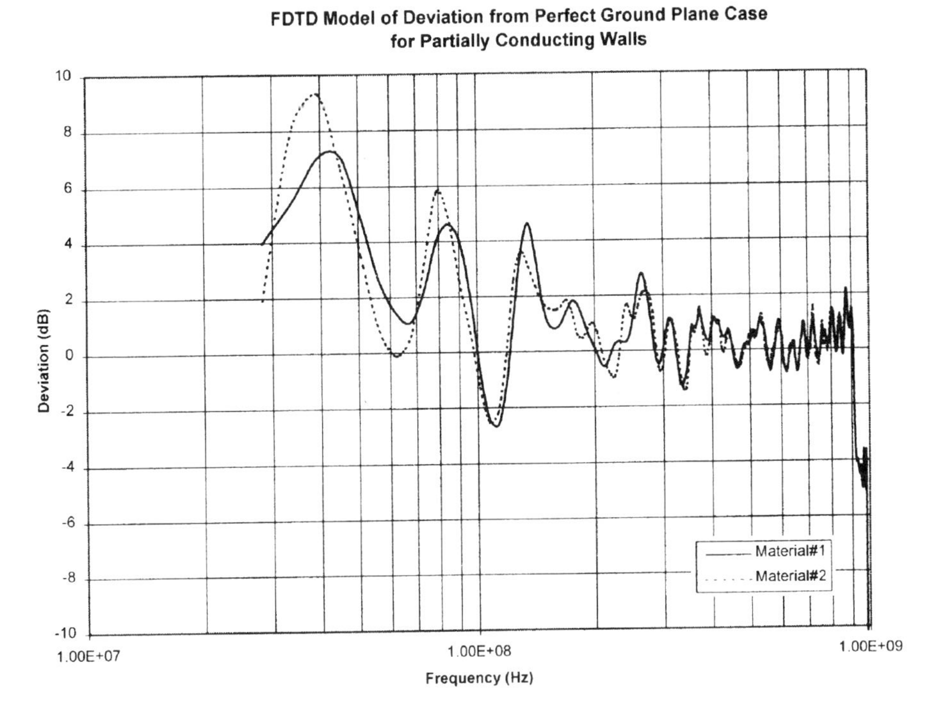

Figure 8.26 FDTD Model Variation Due to Partially Conducting Walls Around OATS

surface patch plane) which must be fine enough to be useful at high frequencies. MoM is particularly useful for problems involving long wires and poles. FDTD requires a fine grid, again for high frequency accuracy, and it is especially useful for problems involving partially conducting walls and intervening dielectrics. Although the analysis shown here was limited in frequency from about 500 to 800 MHz, this could be easily extended by finer wire grids and cell sizes. Since practical OATS calibration has shown most of the problems due to site construction to be apparent at low frequencies rather than high frequencies, finer mesh models may not be necessary.

8.4.2 Semi-Anechoic Shielded Rooms

Semi-anechoic rooms are typically shielded rooms with an RF absorber material on the inside walls and ceiling and are intended to represent an OATS environment, except without the external ambient. The RF absorber material usually works well as an absorber, as long as it is

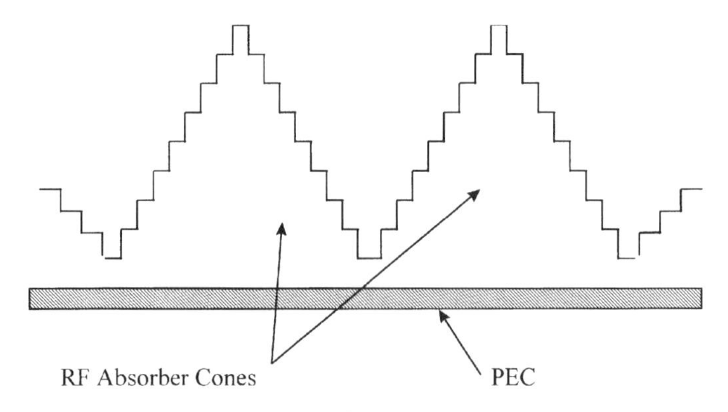

Figure 8.27 Stair Step Example for RF Absorber Cones

thick compared to the wavelength. Naturally, if the absorber is believed to be working as intended, then the semi-anechoic room can be modeled as a perfect OATS.

Often design engineers wish to model the effects of the absorber material and determine the frequency range where the semi-anechoic room will operate like an OATS. This modeling can be performed in a straightforward manner using the FDTD technique.

As stated in earlier chapters, once the material parameters are known, then each FDTD cell can be individually assigned to those parameters. In the case of pyramid-shaped RF absorber cones, the cones can be modeled using the conductivity and permittivity of the cone material in a stair case fashion. Figure 8.27 shows a two-dimensional view of a stair case model for an RF absorber cone. Stair-casing has been shown to be an accurate way to model a non-rectangular feature in FDTD, as long as the step size is about $\frac{1}{20}$th λ (at the highest frequency of interest).

For example, if a shielded room 17 m × 20 m × 5 m high is to be used as a semi-anechoic room, it must be broken into FDTD cells. For a highest frequency of one GHz, the lambda/20 cell size will be 0.016 meters, resulting in over 414 million cells. Since an FDTD cell typically requires about 45 bytes of memory per cell,[8] this results in about 18.5 Gbytes of RAM in the computer to avoid page swapping.[9]

[8]The exact number of bytes per FDTD cell is dependent on the exact code implementation of the FDTD technique.

[9]Page swapping occurs when the program must swap pages of memory in and out of the RAM. Page swapping slows the effective computer speed tremendously, and should be avoided.

Although this is a considerable requirement, the number of time steps in the FDTD simulation will still be greater than if a smaller number of cells were used, and such memory-efficient models should be considered.

In most cases, the RF absorber is a more efficient absorber at higher frequencies. For this example of a semi-anechoic room, the absorber can be assumed to be providing sufficient absorption at higher frequencies, but some estimation of its performance must be made for lower frequencies. If the maximum frequency of interest is set to 300 MHz, the total number of cells is reduced to 13.6 million or 612 Mbytes of RAM. At a 100 MHz maximum frequency, the total number of cells is further reduced to .41 million or 18.5 Mbytes of RAM.

Another more convenient method can be used to model the absorber material using progressive layering. The cones are replaced in the model with a number of layers of material where the parameters are varied to match the effects of the original cones. The material parameters are adjusted depending upon the percentage of the layer that is made up from the cone. Figure 8.28 shows an example of the layers. One key advantage to layering is the requirement for $\lambda/20$ cell sizes for stair-stepping is eliminated, and 'normal' sized cells can be used, saving additional computer memory and time. With layering, and using the $\frac{1}{10}$th λ cell size, the number of cells for a 300 MHz maximum is reduced to 1.7 million cells or 76.5 Mbytes of RAM.

The use of ferrite tiles between the shielded room wall and the

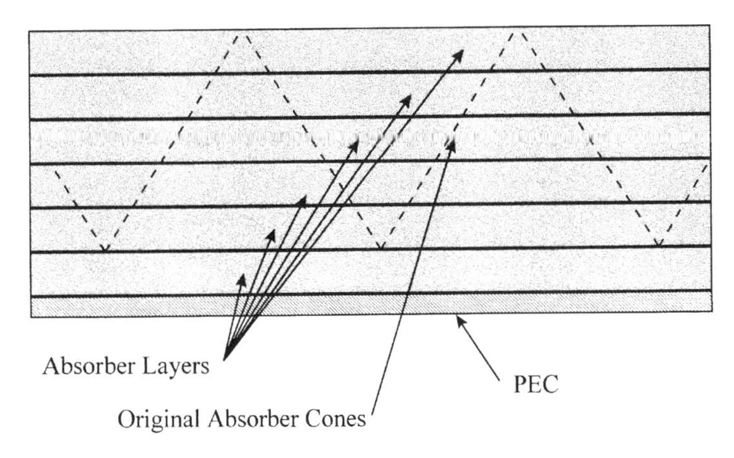

Figure 8.28 Equivalent Layer Example for RF Absorber Cones

absorber cones can be modeled as well. Modeling ferrite materials in FDTD is more difficult than that of most materials, since the constitutive parameters of the material change with frequency. Therefore ferrite material must be modeled using one of two techniques; the FDTD equations must be modified to include a time domain convolution for each cell in the ferrite, or the constitutive parameters must be approximated over frequency bands, and the final result assembled from the results in each frequency band.

8.4.3 GTEM Cells

GTEM cells are commonly used for radiated susceptibility of small devices, and may also be used for radiated emissions. A GTEM cell is a pyramidal shaped metal chamber with an internal septum as shown in Figure 8.29. The GTEM cell is typically loaded with RF absorber material at the far end, and the septum is terminated with a resistive load. When used for susceptibility, the GTEM cell is specified to provide a uniform field level over a specified volume within the cell.

Since there is absorber material and a septum (where the induced RF currents are different on either side of the metal septum), FDTD is the preferred modeling technique. A stair-stepping technique is required to correctly define the GTEM walls, and the layering technique described earlier can be used for the absorber material.

Since FDTD is a time domain technique, a wide range of frequencies can be analyzed with one simulation. Monitor points can be placed

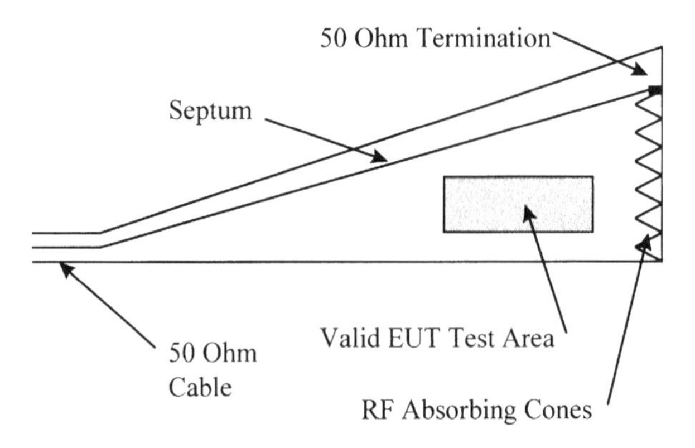

Figure 8.29 GTEM Test Chamber Diagram

over the entire specified volume, and analyzed to ensure the field strength is consistent within the limits required.

8.4.4 Mode-Stirred Chambers

Mode-stirred chambers are high Q shielded rooms with a large metal paddle wheel. The paddle wheel is rotated through various positions to change the standing wave pattern in the shielded room, thus 'stirring' the 'modes'. Mode-stirred chambers are typically used for radiated susceptibility where a given field strength is desired over a specified volume in the chamber and over a given frequency range. Correct operation of mode-stirred chambers require a high Q environment and operate only at frequencies where the wavelength is short compared to the chamber dimensions.

Since the RF currents induced on either side of the metal paddle wheel will be different, a volume-based modeling technique is required. FDTD is again convenient since it provides a wide frequency response with a single simulation. The source is positioned to provide no direct stimulation of the target area, and the peak field over the specified volume area is monitored over all the possible positions of the paddle wheel. The paddle wheel is rotated a fixed number of degrees, often only two to three degrees, and the electric field over the target area is noted. Ideally, after all the various rotational positions have been used, the peak electric field values at each target position will then be the same. Therefore, a number of FDTD runs will be needed to simulate a mode-stirred chamber in operation, one for each paddle wheel position. The peak electric field strength at each monitor point in the target area must be found at each frequency for all paddle positions. Figure 8.30 shows a two-dimensional view of a typical mode-stirred chamber.

Another important parameter of the mode-stirred chamber is the chamber Q. It has been found that nonconductive materials can have a significant effect on this Q, and must be included in the FDTD model.

8.4.5 Section Summary

The choice of modeling for a variety of different EMI/EMC test site applications was demonstrated in this section. The use of modeling technique was shown to depend upon the situation and environment to be modeled. Effects of the measurement environment can be very

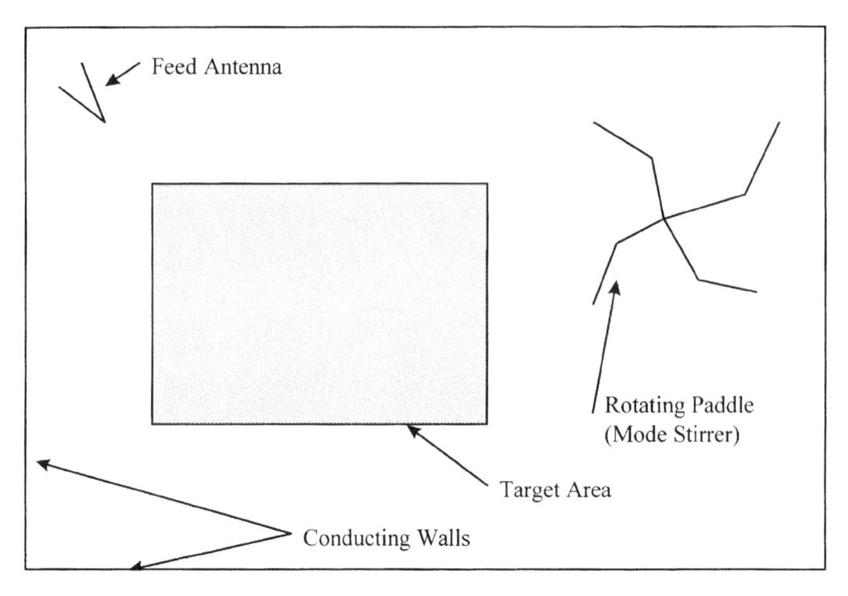

Figure 8.30 Mode-Stirred Chamber Diagram

important when comparing modeled and measured results and must be included when appropriate.

8.5 Antennas

Most of the discussions so far have assumed a perfect antenna. That is, the antennas had no resonant frequency or direction. For example, in FDTD the observation points are receive "antennas" that are completely isotropic and can measure the x, y, and z components of the electric and magnetic fields separately. This is impossible to achieve in practice, and therefore modeling the receive antenna itself can be important when attempting to compare modeled and measured results. (See Chapter 9 on Validation.)

When modeling antennas, it is important to include the source impedance. The real world antenna will have a radiation impedance and the amount of power radiated is dependent on the combination of the source and the radiation impedance.

8.5.1 Dipole Antennas

Dipole antennas are fairly straightforward to model. Since they are long, thin wire structures, they are ideal for modeling in the MoM technique. An odd number of segments are selected along the entire dipole to allow the center segment to be the source with the appropriate source impedance. The segment size must be small compared to the wavelength to provide the accuracy desired. Chapter 7 presents further discussion on a 30-MHz tuned dipole antenna. More complicated antenna structures, such as Biconical antennas or end-loaded dipoles, can be easily modeled using a series of wires properly combined.

8.5.2 Horn Antennas

Horn antennas could be modeled using an MoM approach (either surface segments or a wire frame grid approach) but the basic limitation of most MoM techniques is the currents on either side of the patch or segment must be the same. This is not true for real world horn antennas where the surface currents on the inside of the horn will be very different than those on the outside. A model of a horn antenna using the MoM might be accurate enough if only straight boresight field strengths are required. However, if off-boresight field strengths are desired, a volume based technique such as FDTD will be required.

Since horn antennas have surfaces which are not parallel to the x-, y-, and z-planes, one or more of the antenna's surfaces must be created using stair stepping. Stair steps of $\frac{1}{20}$th λ have been shown to give good accuracy for most applications.

FDTD can provide animation views of the fields as they propagate through the horn antenna. The diffraction at the horn's edge, and the reflection due to the discontinuity at the edge can be easily seen during these animations. If far field patterns are desired, the field extension technique described in Chapter 3 might be needed to avoid excessively large FDTD computational domains.

8.5.3 Effects of the Ground Plane on the Antenna Factor

The earlier section on the EMI/EMC test sites described that all measurements be performed over a perfect ground plane. When using real world antennas, the conversion between electric field strength and terminal voltage is through a far-field, free-space, antenna correction factor.

Modeling can be used to determine the effect of the ground plane on these free-space antenna correction factors. The antenna can be polarized either horizontally or vertically relative to the ground plane, and then moved closer while the antenna correction factors are calculated. In the vertical antenna case, the ground plane will have a significant effect. This can be seen by placing a metal plane at various distances from the end of a dipole (or other style) antenna that is electrically large, but small enough to not introduce reflections between the source and the dipole antenna.

8.5.4 Effects on Antenna Radiation When Placed inside a Shielded Enclosure

It is a common practice to make shielding effectiveness measurements on metal shielded enclosures by first placing an antenna in free space and then placing the antenna inside a shielded enclosure. The difference in radiated electric field is then considered to be the shielding effectiveness of the enclosure.

While this test is generally understood to have serious flaws due to the change in the antenna's radiation characteristics when placed inside a shielded enclosure, these changes are seldom taken into account during the measurement process. However, these effects could be examined, and possibly removed, using modeling. The change in the radiation impedance will have an effect on the amount of power radiated from the antenna. Each test setup could be modeled and the radiation impedance determined.

Since we are not trying to determine the shielding effectiveness of the enclosure through modeling, the fields outside the enclosure are not important for this example, and the structure could be modeled using MoM or FDTD. A simple loop antenna or monopole antenna could be placed in free space and then inside a metal box while the antenna radiation impedance is found for a frequency range. These results will show dramatic differences between the radiation impedances and, as a result, dramatic differences in the amount of power radiated. It is possible that results such as these will discourage people from using this test technique to determine absolute shielding effectiveness.

8.5.5 Section Summary

A variety of antenna related topics were examined in this section for use with modeling techniques. The effects of real world antennas can be added into the model's normally perfect antennas either as a "second stage" in a multiple stage approach, or could be included in the initial model directly.

8.6 Summary

This chapter covers a wide range of modeling topics of interest to EMI/EMC engineers. Multiple stage models can be used when the various sections are electromagnetically separable, and the modeling techniques can be varied to allow each individual stage to be optimized. Test sites can be evaluated before construction, to show the effects of changes to "typical" recommendations. Antennas and other measuring devices can be modeled to evaluate them or to allow their effects to be included into the overall results. The limitations of the application of the various modeling techniques is dependent on the user's imagination.

References

1. B.R Archambeault, "Modeling of EMI Emissions from Microstrip Structures with Imperfect Reference Planes," *Applied Computational Electromagnetics Society Symposium*, March 1997, pp. 1058–1063.
2. S. Poh, "Electromagnetic Emissions from Microstrip Discontinuities in Interchip Packaging," Internal Digital Equipment Corporation Report, 13 Feb, 1991.
3. S. Daijavad, H, Heeb, "On the Effectiveness of Decoupling Capacitors in Reducing Electromagnetic Radiation from PCBs," *1993 IEEE International Symposium on Electromagnetic Compatibility*, pp. 330–333.
4. B.R. Archambeault, C. Brench, "Shielded Air Vent Designs Guidelines from EMI Modeling," *1993 IEEE International Symposium on Electromagnetic Compatibility*, pp. 195–199.
5. D.M. Hockanson, J.L. Drewniak, T.H. Hubing, and T.P. VanDoren, "Application of the Finite Difference Time Domain Method to Radiation from Shielded Enclosures," *1994 IEEE International Symposium on Electromagnetic Compatibility*, pp. 83–88.
6. B.L. Brench, C.E. Brench, "Shield Degradation in the Presence of External Conductors," *1994 IEEE International Symposium on Electromagnetic Compatibility*, pp. 269–273.
7. B.R. Archambeault, C.E. Brench, "Modeled and Measured Results from Two Proposed Standard EMI Modeling Problems," *1995 IEEE International Symposium on Electromagnetic Compatibility*, pp. 349–352.

8. B.R. Archambeault and K. Chamberlin, "Modeling the EMI Performance of Various Seam Shapes," *Applied Computational Electromagnetics Society International Symposium*, 1995.
9. J.L. Drewniak, T.H. Hubing, T.P. Van Doren, "Investigation of Fundamental Mechanisms of Common-Mode Radiation from Printed Circuit Boards with Attached Cables, Internal Report, University of Missouri-Rolla.
10. T.H. Hubing and J.K. Kaufman, "Modeling the Electromagnetic Radiation from Electrically Small Table-Top Products," *IEEE Transactions on Electromagnetic Compatibility*, Vol. 31, pp. 74–84, 1989.
11. P. Harms, R. Mittra, and J. Nadolny, "Simulating Measurements for a Cable Radiation Study," *IEEE Transactions on Electromagnetic Compatibility*, Vol. 38, pp. 25–31, 1996.
12. J.D. Kraus, *Antennas*, McGraw-Hill Book, New York, 2nd Ed., 1988.

9

Model Validation

9.0 Introduction

It is never enough to simply "believe" a model has created correct answers; engineers must be able to prove it or validate the model. Simulation is usually performed on a computer, and the old phrase Garbage In, Garbage Out (GIGO) still applies. There are different levels of validation, and care should be taken to understand which level is necessary for a given problem. Also, not every problem needs to be independently validated. If every simulation must be validated with measurements or calculations, it would be unnecessary to perform any simulations at all.

9.1 Computational Technique Validation

One level of model validation is the computational technique validation. This is usually unnecessary in most EMI/EMC modeling problems, since the computational technique will have been validated in the past by countless others. If a new technique is developed, it too must undergo extensive validation to determine its limitations, strengths, and accuracy but, if a "standard" technique such as Finite-Difference Time-Domain (FDTD), Method of Moments (MoM), and Finite Element Method (FEM), is used, the engineer need not repeat the basic technique validation. This is not to say, however, that incorrect results will not result if an incorrect model is created or if a modeling technique is used incorrectly.

Once the basic technique is validated, the way it is to be used must still be validated. For example, if the MoM is to be used on a simple dipole antenna, a common rule of thumb is to use wire segments of about $1/10$th λ at the highest frequency of interest. Is this accurate enough? Naturally, it depends upon the application and the accuracy requirement. For many applications $1/6$th λ might be adequate, while for others, $1/20$th λ might be required. There is a point of diminishing returns with the size of the segments. In fact, there is even a risk of creating errors with excessively small segments (due to computational numerical noise). Figure 9.1 shows an example of the maximum field strength from a half-wave dipole antenna as the segment size is varied. A convergence point is eventually reached, and the amount of allowable error from the convergence value is used to determine the required segment size.

A similar case can be made for FDTD. A typical rule of thumb is to ensure the cell size is no more than $1/10$th λ on a side. For many applications, much larger cells provide sufficient accuracy, while for other applications much finer cell resolution is required.

9.2 Individual Software Code Implementation Validation

The next level of validation is to ensure the software implementation of the modeling technique is correct, and creates correct results. Naturally, everyone who creates software intends it to create correct results; however, it is usually prudent to test individual codes against the types of problems for which they'll be used.

For example, a software vendor will have a number of different examples where their software code has been used, and where tests or calculations have shown good correlation with the modeled results.

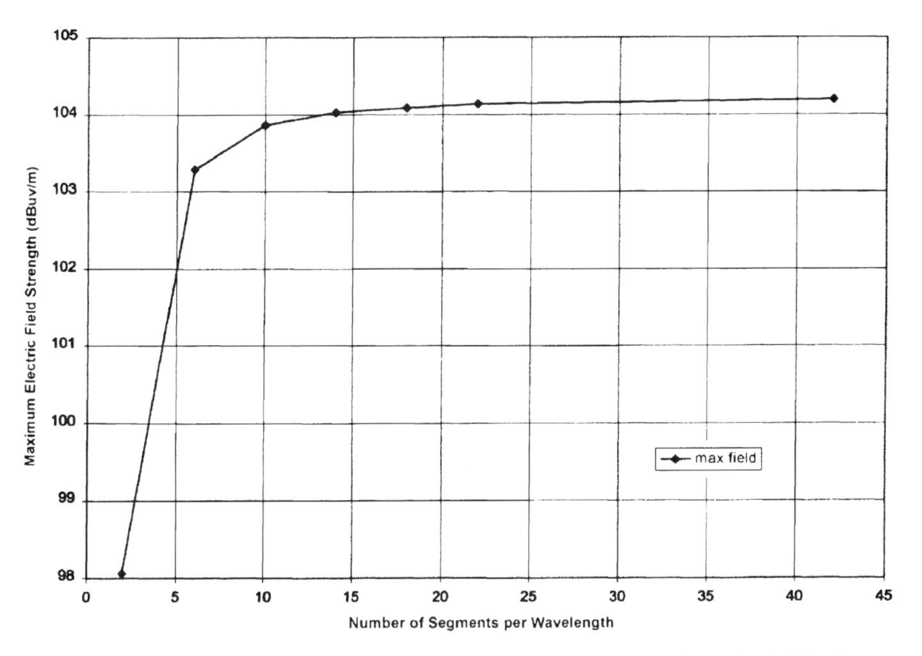

Figure 9.1 Effect of MoM Segment Size on Maximum Field Strength for a Simple Dipole

This is good, and helps the potential user to have confidence in that software code for those applications where there is good correlation. However, this does not necessarily mean that the software code can be used for any type of application and still produce correct results. There could be limitations in the basic technique used in this software, or there could be difficulties in the software implementation of that specific problem.

As a first example, consider the case of a software code using the FEM. As shown in Chapter 5, this method is very useful for certain applications, especially where the geometry is nonrectangular, or where the fields are changing very rapidly. So a typical software code might have validation examples where the software was used to predict the fields within a waveguide, or the fields within a resonant cavity, or even the cross-coupling between adjacent traces on a PCB. However, if the problem required the modeling of an unshielded circuit board with a long wire attached, the computational technique would be used for an application where it is not well suited. Validation of the basic computational technique is still valid, as is the applications where the software code was validated. However, the specific class of problems that include long attached wires are often difficult to properly model in FEM models, and all the other validations would not have shown this validation problem.

As a second example, a simple shielded box with a long wire attached is to be modeled. The source for this model is the common-mode voltage between the shielded box and the long attached wire (possibly known from a different model, or from measurements). Since there is a long wire involved, then the MoM would most likely be selected. MoM allows both surface patches (useful for modeling the shielded box) and wire segments (useful for modeling the long attached wire). A software code in common use today is a well-known implementation of the MoM and could easily be applied to this problem. This particular code has been validated by many engineers in many different applications, as well as its authors. The ample number of validation reports might lull an engineer into complacency (as long as the model does not violate the 'rules' of MoM). However, in this example, the long wire must be attached to the surface patch, and this area of contact will have very high RF currents. In this particular implementation of MoM, substantial errors occur when high currents are present at wire segment to surface patch interfaces. The validation of the basic computational technique (MoM) is still valid, as is the application in which the software code was validated. However, the specific class of problems that include long attached wires to surface patches were not included in the validation examples (for good reasons), and the other examples of validation do not apply to this example. As a further note, in this case, if the shielded box had been modeled as a wire grid frame box, and the intersection point of the long wire modeled to allow the currents on the shielded box to flow radial in all directions, correct results could be obtained.

So when a previous validation effort is to be extended to a current use, the types of problems that have been validated in the past must closely match the important features of the current model.

9.3 Model Validation Using Measurements

The most common type of validation for EMI/EMC applications will be actual measurements. This is largely due to the fact that EMI/EMC type problems do not lend themselves to closed form calculations. It is obvious, but often overlooked, that the same problem must be used in both the modeling and the measurement cases. All important features must be included in both. It is unlikely that a case where a simple OATS was used to measure the emissions from a printed circuit board would validate a model of the PCB only. The OATS ground plane would have an enormous effect on the measurements and, if not included in the model, the results are very unlikely to agree. Figure 9.2 shows the deviation due to the OATS ground plane and receive antenna scan for a simple horizontal dipole source. The effect of the ground plane can be ± 6dB due to constructive and destructive interference. Once frequencies higher than about 200 MHz are reached, the receive antenna scan from one to four meters ensures a constructive interference effect with the corresponding +6 dB deviation between the case with no ground plane and with the ground plane.

Another consideration to model validation by measurement is the accuracy of the measurement itself. While most engineers take great comfort in data from measurements, the repeatability of these measurements (either from test laboratory to test laboratory, or even within the

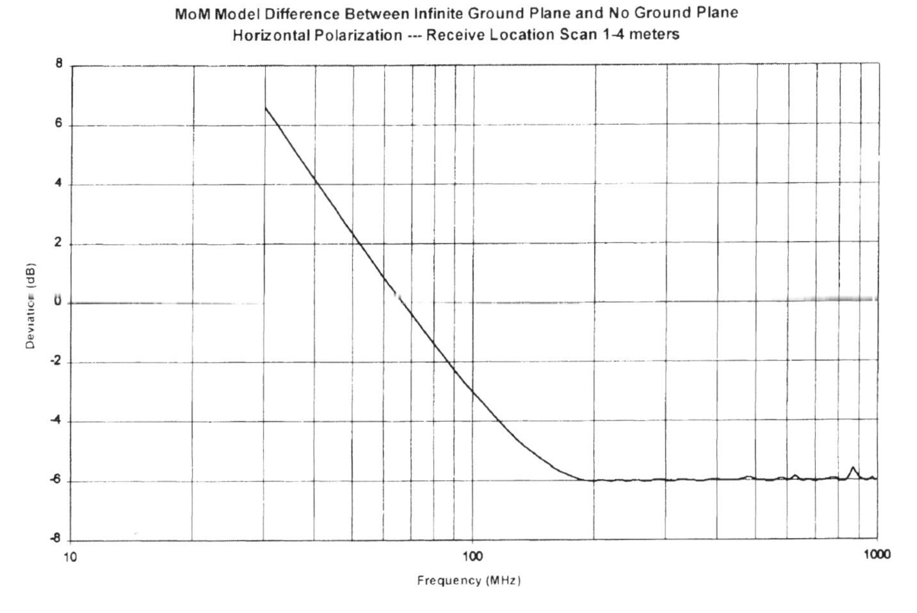

Figure 9.2 MoM Modeled Difference for OATS With and Without Infinite Ground Plane

same test laboratory on different days) in a commercial EMI/EMC test laboratory can easily be unrepeatable within ±6 dB.[1] Military EMI/EMC laboratories are considered to have a much higher measurement uncertainty due to the plain shielded room test environment. So the test environment's repeatability, accuracy, and measurement uncertainty must be included when evaluating a numerical model's result against a measurement. The agreement between the modeled data and the measurement data can be no better than the test laboratory's uncertainty. If measurement data disagrees with modeled data, then some consideration should be given to the possibility that the measurement was incorrect and the model data correct. Therefore it is essential to avoid measurement bias and equally consider both results as correct. When two different techniques provide different results, all that can be logically known is that one of them is wrong.

The required test environment varies depending upon the EMI/EMC standards. Some EMI/EMC measurements are performed over a ground plane. Some require all cables to be placed in an given manner, while others might require the cables be moved to maximize the emissions. Some MIL-STD measurements are performed with the equipment placed on top of a metal table, while most commercial standards require the tabletop devices be placed on a wooden table, and floor standing units be placed on the metal ground plane (and isolated). It is well known and understood that measurements of the same equipment in different type EMI/EMC test labs will result in different measured levels. Therefore, the model results should be expected to be different if the test environment is not carefully matched.

A true example of the potential error when comparing modeled results with measurement results occurred for a simple circuit board with a short microstrip trace on it, shown in Figure 9.3a. All the physical dimensions were carefully reproduced in the model. The test environment was a semi-anechoic chamber, good enough to be considered an Open Area Test Site (OATS), and the orientation of the PCB in relation to the receive antenna carefully noted. However, the initial results were surprisingly different between the model and the measurements. Upon reexamination of the measurement process, it was noted that the microstrip on the PCB was fed from below with a coax cable

[1]The poor measurement accuracy (or repeatability) is due to measurement equipment, antenna factors, site measurement reflection errors, and cable movement optimization.

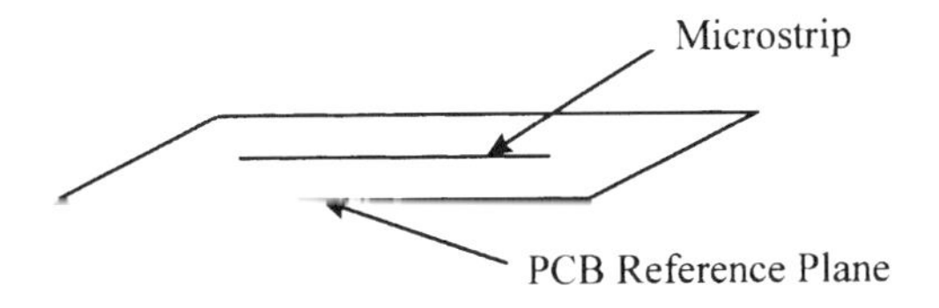

Figure 9.3a Initial Model of Microstrip on a Printed Circuit Board

which led straight down from the center of the PCB ground plane to the floor of the chamber (Figure 9.3b). This extra conductor was certainly part of the physical measurement test, and once it was removed from the test environment by placing ferrite along the shielded cable to remove any unwanted RF currents, the results matched closely. Figure 9.4 shows the difference between the cases with and without the vertical cable.

An extension to the above example would include the measurement scan locations. While it is quite easy to instruct the computational

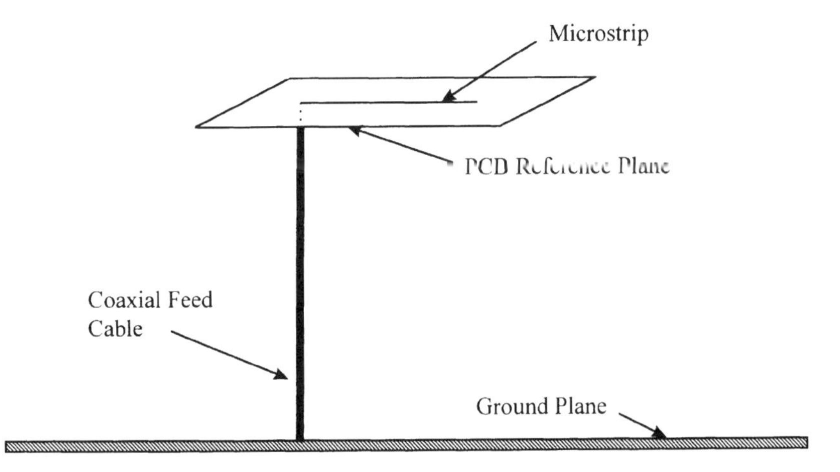

Figure 9.3b Initial Validation Measurement Setup for Microstrip on a Printed Circuit Board

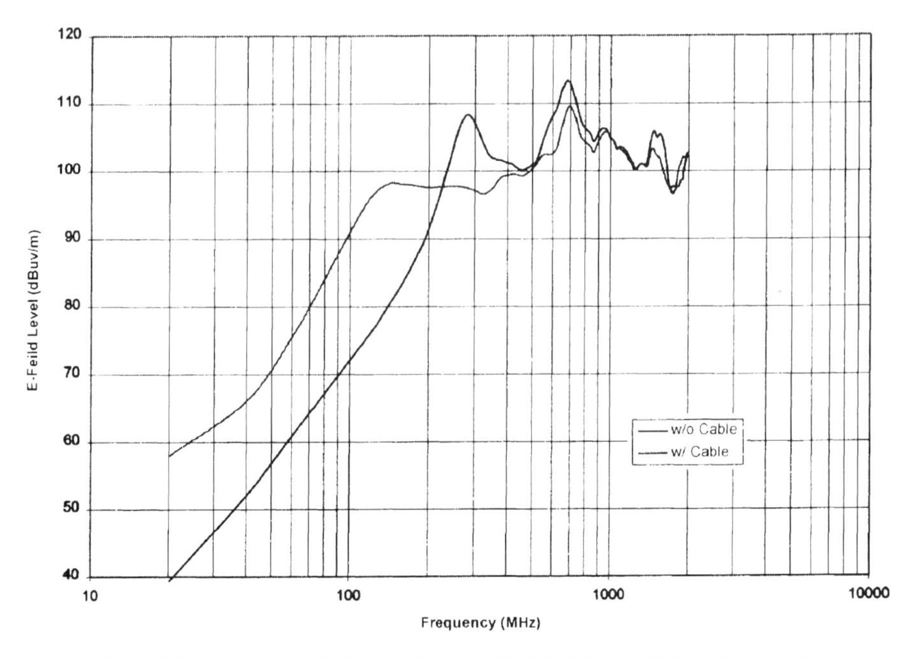

Figure 9.4 Electric Field Difference Between Models With and Without Feed Cable

model to report the maximum field strength *in any direction*, this is not the physical scan area used in any laboratory environments. Commercial EMI/EMC test laboratories measure at three or ten meters distance and scan the receive antenna height from 1 to 4 meters. This corresponds to a cylinder-shaped receive location area, as shown in Figure 9.5. For the above case of the microstrip on a printed circuit board, the maximum emissions would be directed upward. If the computational model was set up to determine the maximum emission in any direction, and the measurement made over the normal cylinder area, then the results are not likely to agree, as shown in Figure 9.6.

So before comparing computational model results to a measured result, care must be taken to ensure that the two results are comparable; that is, they must both represent the same physical environment.

9.4 EMI/EMC Model Validation Using Intermediate Results

Computational modeling provides a tremendous advantage over measurements since physical parameters may be viewed in the computa-

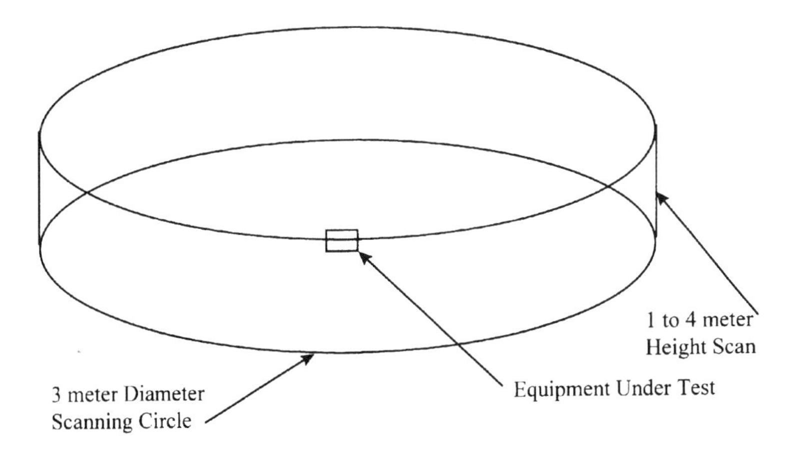

Figure 9.5 Real-World EMI Antenna Receive Location Scan Area

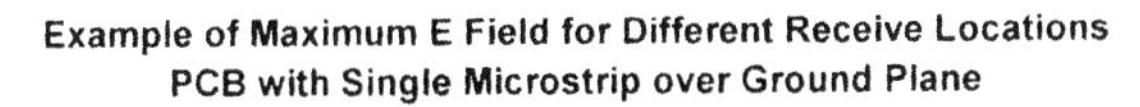

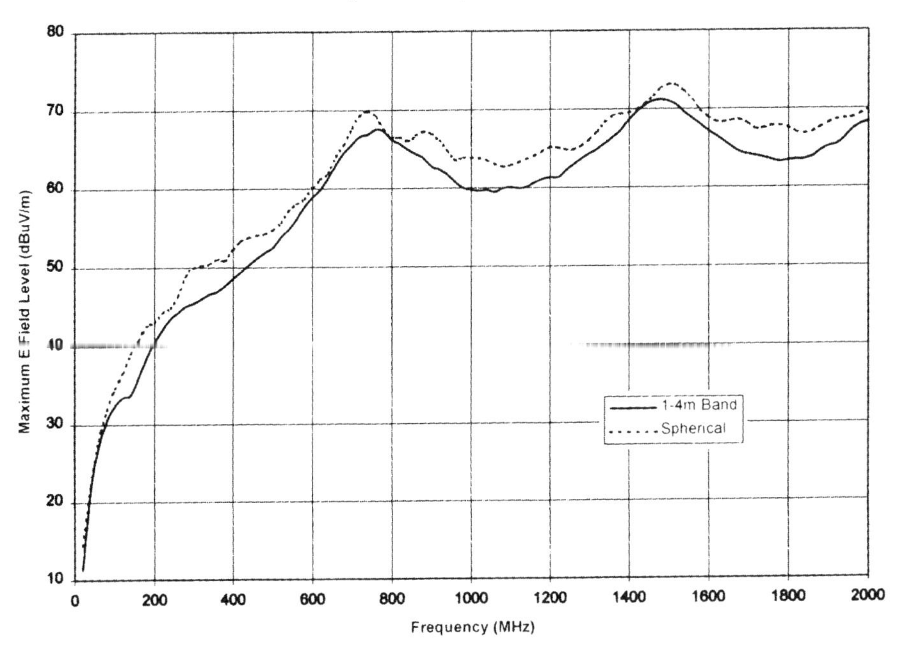

Figure 9.6 Difference in Received Maximum Electric Field Due to Scan Locations

tional model where they could never be physically viewed in the real world. Electric fields, magnetic fields, and RF currents on a plane can all be viewed within the computational model, but can not be viewed directly in the measurement laboratory.

These parameters are used as an intermediate result within the computational model, and can be very useful to help validate the model has performed correctly. While the final far-field result may be the goal of the simulation, the intermediate results should be examined to ensure the model was operating as theory, experience, and intuition require.

9.4.1 RF Currents on a Conducting Surface

When using MoM, as seen in Chapter 4, the currents over the entire structure are found and then the radiated electric fields are found from the RF currents. These currents provide significant insight to the computational results' validity.

Figure 9.7 shows an MoM model of a simple microstrip over an imperfect reference plane where part of the reference plane under the microstrip is missing. The return current in the reference plane would be expected to mostly travel under the microstrip until the void area, and then go around the void area, until it could again travel under the microstrip. If all the MoM patches are not properly electrically connected, and an imperceptible void exists in the area where no void

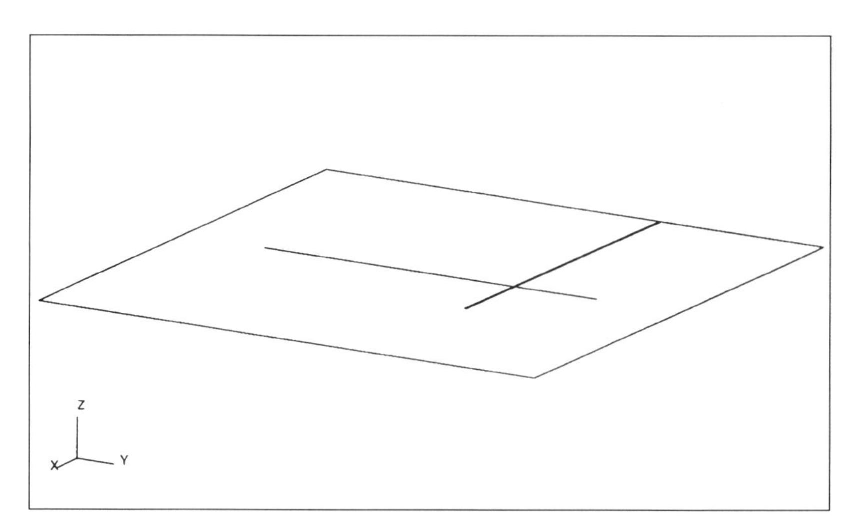

Figure 9.7 Example of Microstrip Trace Over Reference Plane with Split

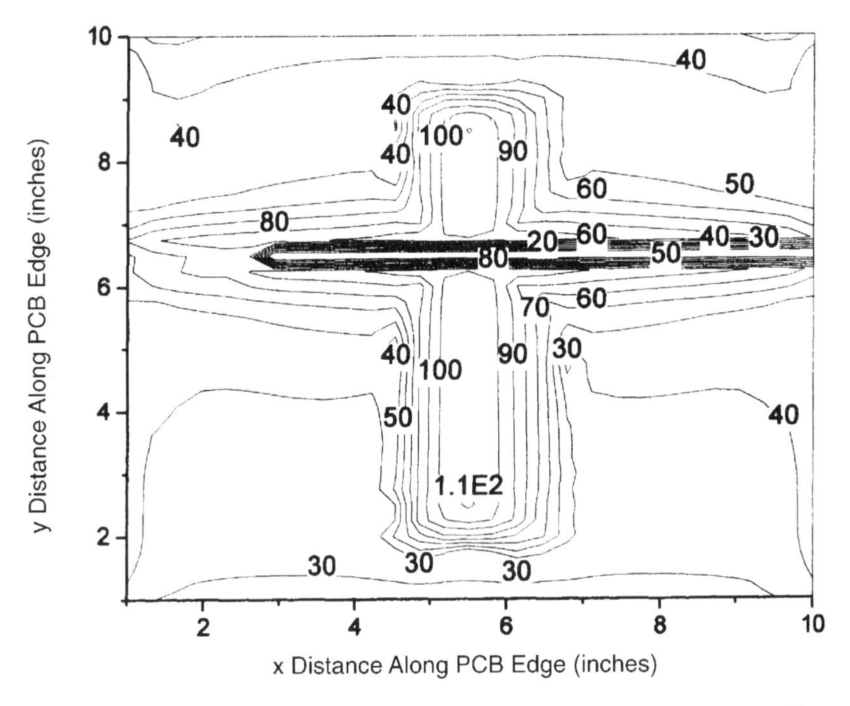

Figure 9.8 Contour Plot of Incorrect RF Current at 100 MHz (dBuA) on Split Reference Plane Due to Microstrip (Due to Incorrect Model)

was intended, then the currents might behave very differently than expected. Figure 9.8 shows an example of the incorrect RF return currents for this case. The microstrip area runs vertically in this figure and is the area of the highest current amplitude, as expected. However, something 'funny' or unexpected is clearly indicated in Figure 9.8, since the currents do not seem to crowd around the end of the void as would be expected, and further investigation is required. If only the far-field radiated electric field were observed, although the result would be incorrect, there would be no apparent indication that there was anything wrong. Figure 9.9 shows the expected RF current crowding around the end of the void, once the incorrect model was repaired.

Another indication of potential problems is when the RF current changes too rapidly from wire segment to wire segment or surface patch to surface patch. The wire segment size or patch size is selected to be electrically short to ensure the currents do not vary rapidly from segment to segment. Therefore, it is recommended that the RF current

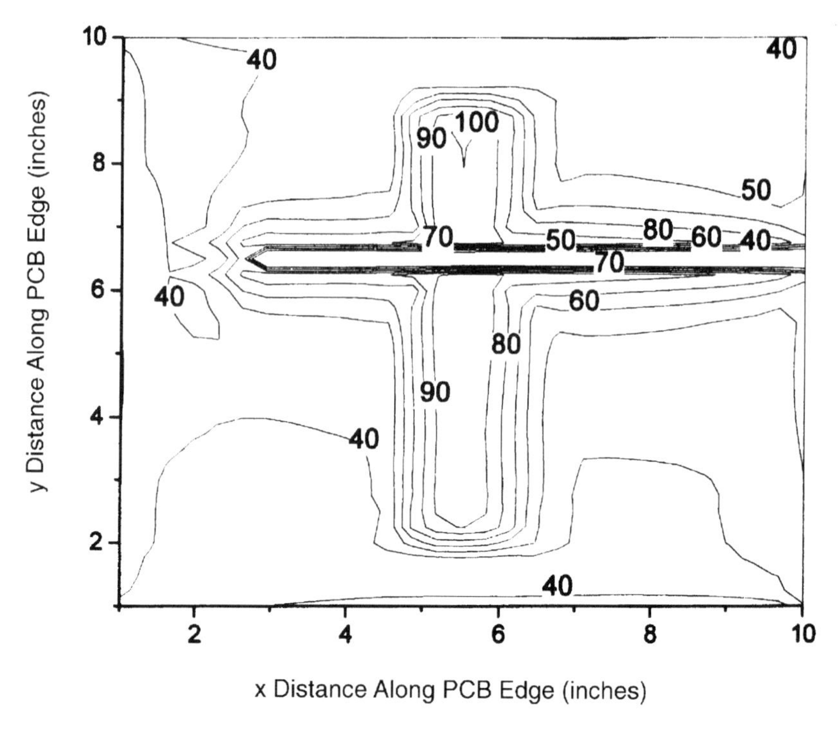

Figure 9.9 Contour Plot of Correct (Expected) RF Current at 100 MHz (dBuA) on Split Reference Plane Due to Microstrip

distribution be examined to ensure the expected slowly varying distribution is present.

When wires are not terminated into another wire or structure, the RF current must go to zero at the unterminated end. However, if the segment size is too large, the current cannot smoothly go to zero, and errors can result. Therefore, it is recommended that the RF current distribution be examined to ensure the current goes to zero at the ends of any unterminated wires.

9.4.2 *Animated Electric Fields*

When using FDTD the fields are found for all the cells within the computational domain for each time step. Typically, the final result desired is the field strength at a specific location or number of locations. However, viewing the electric field as it propagates through the compu-

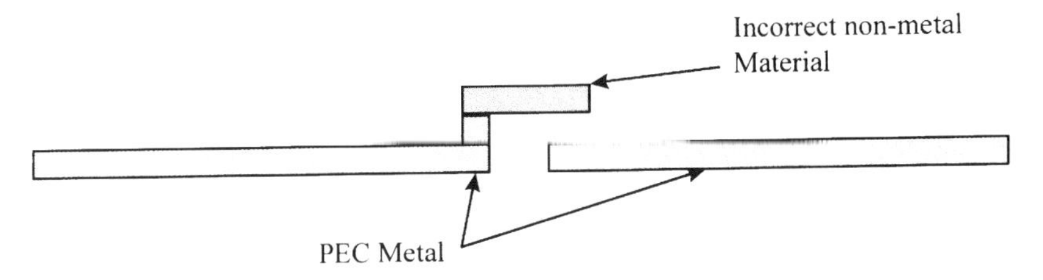

Figure 9.10 FDTD Seam Model Example With Error in Material Assignments

tational domain can provide significant insight to the computational result's validity.

Figure 9.10 shows an example of a simple seam along two mating edges of a shielded cover. A wire source is placed nearby and excited with a pulse. The metal of the covers is assumed to be perfectly conducting, that is, the majority of leakage is due to the opening in the seam and not direct conduction through the metal itself.

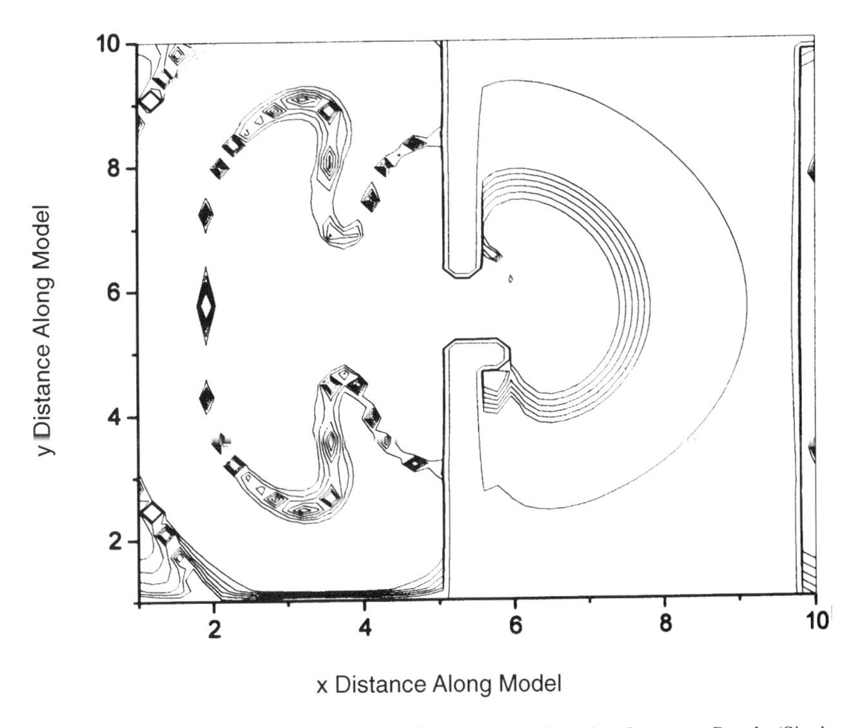

Figure 9.11 FDTD Electric Field Contour Plot of Seam Showing Improper Result (Single Animation Frame)

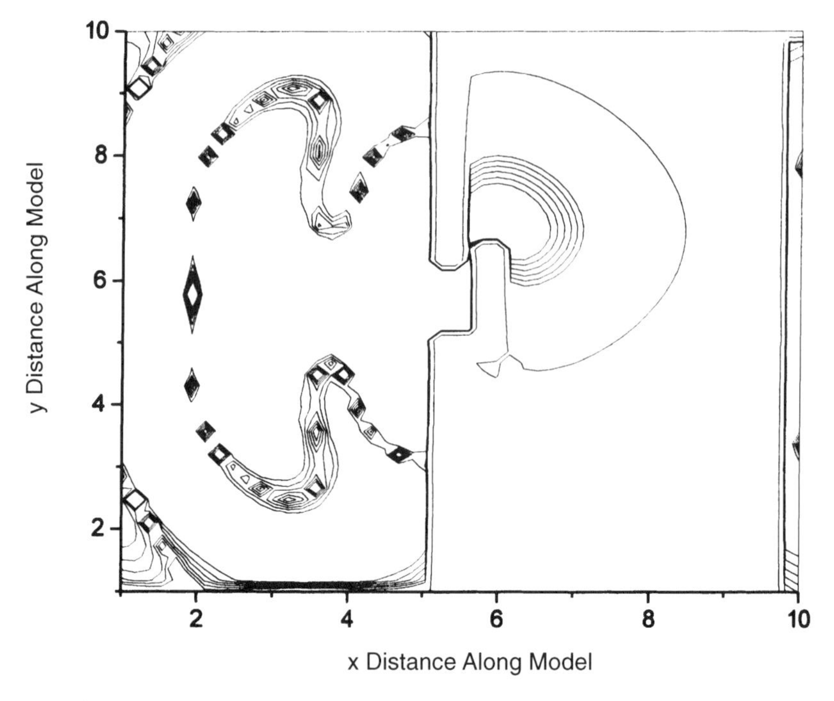

Figure 9.12 FDTD Electric Field Contour Plot of Seam Showing Expected Result (Single Animation Frame)

An error when creating the initial model resulted in part of the seam to be made of a material with a relative dialectic of 1.0 (air) instead of Perfect Electrical Conductor (PEC) metal. A contour plot view of a single frame of the electric field animation in Figure 9.11 shows the electric field radiating uniformly to the right side, indicating the overlapping seam of metal was not correctly modeled. Once the model was corrected (i.e., PEC metal restored to all intended locations) the model was re-run with the correct view of the electric field (Figure 9.12). The expected direction of the radiated electric field, due to the presence of the overlapping seam, is now present.

9.5 Summary

This chapter has discussed EMI/EMC modeling validation using a number of different techniques, depending upon which is most appropriate. Engineers should validate their models to ensure the model's correctness, and to help understand the basic physics behind the model.

Measurements can be used to validate modeling results but extreme care must be used to ensure the model correctly simulates the measured situation. Omitting feed cables, shielding or ground reflections, or different measurement scan areas can dramatically change the results. An incorrect model result might be indicated when, in fact, the two results are obtained for different situations, and should not be directly compared.

Intermediate results can also be used to help increase the confidence in a model. Using the RF current distribution in an MoM model, or the animation in an FDTD model, can help ensure the overall results are correct by ensuring the intermediate results are correct. These intermediate results have the added benefit of helping the engineer better understand the underlying causes and effects of the overall problem.

10

Standard EMI/EMC Problems for Software Evaluation

10.1 Introduction

When considering the acquisition of an EMI/EMC modeling tool, there are many things to consider. The most obvious of these is whether the tool can be used to model the particular types of problems at hand. Additionally it is necessary to understand the graphical user interface (GUI) and the input and output file formats. The file formats may be of particular importance if it is intended to link other tools, such as circuit simulation or mechanical CAD, to the EMI/EMC tools.

Most vendors are more than happy to demonstrate their GUI, so there is little need to go into great detail of what options are generally available. The key characteristics of the GUI are that it permits the rapid and complete entry of the problem and that it clearly and conveniently displays the results of the simulation.

The input and output (I/O) file formats can take many forms, from simple ASCII to proprietary. If there are special requirements for interfaces to other tools, it is essential to understand how this can be achieved and to see practical examples of the converted files. Depending on the use, it may be important to have the capability to manipulate either the input or output files directly. This ensures that the interface remains suitable as the EMI/EMC tools with which it interfaces are updated.

The real test of suitability is whether the particular implementation of the computational EM technique used in the tool can solve the class(es) of problems that must be addressed. How the wide range of EMI/EMC problems may be classified can be seen in Chapter 6. The ability to evaluate computer tools reliably for EMI/EMC modeling becomes more important as their number grows. The goal of using standard problems is to permit users to determine a tool's suitability and even to benchmark the tools against each other for real problems. This chapter provides a set of standard problems that can be used as a starting point for the evaluation of the electromagnetic solver. These problems can be added to or replaced as model goals change and as the state of the art progresses.

10.2 General Principles

There are three principles that can guide the creation of a standard model suitable for tool evaluation.

1. Avoid idealized cases.
2. Make the problems reflect reality.
3. Ensure that the tool will do all required of it.

Idealized cases do not provide a good test for EMI/EMC modeling tools, as these are often the most easily modeled. If a vendor's example includes a perfectly symmetrical transmission line or other device, it is important to find out what happens when a small asymmetry is added. If only a narrow range of frequencies is modeled, it is important to find out what happens for wider ranges.

The goal is to obtain a tool from a vendor that is adequate to solve all the EMI/EMC modeling problems that must be addressed by the user. In reality, practical problems are always complex. The usual manner of addressing these problems is to split them into manageable portions and to tackle them piece by piece. The standard problems presented here are representative of how problems are often broken up. However, when dealing with large problems, it can be very illuminating to see how the vendors would split a specific problem to best match their software, rather than providing the problem already split.

It is common for software modeling tools to provide the capability to model just one or two parts of the broken up problem and often to completely ignore the other parts. Depending upon the model objective, the need may be for a tool to solve only one class of EMI/EMC problem, for example, emissions from printed circuit modules. If only one class of problem exists, obviously this becomes the focus of the investigation. However, if the circuit module is installed in a shielded enclosure and has long wires connected to it, there is a much wider range of problems to be solved. Whenever a wide range of problems needs to be addressed, it is possible that more than one modeling approach is needed. In these cases, it is important to be able to smoothly pass data from one component of the tool to another.

In general, when preparing to perform EMI/EMC modeling, the problem must be broken down into the following components: the originating source of the EM energy, the final radiating element, and the coupling mechanism between them. Once these individual components are understood, the most appropriate modeling technique for the task can be selected. The originating source of the EMI signal is usually a high-speed, fast rise time signal. This signal may be present for example, on a bus as a current, or between and the system reference plane and an adjacent module as a voltage. The final radiating element is the "antenna" that causes the emissions to radiate from the equipment under test (EUT). Examples of this radiating element are the wires leaving the EUT with a common mode voltage on them, or a slot or air vent opening. The coupling between the source and the final radiating element can be accomplished through a variety of processes. The coupling could be through the fields internal to the shielded box between the source and a connector/wire leaving the shielded box. Another common coupling mechanism is for currents created in the reference plane by a microstrip bus to travel to another part of the circuit module and be conducted onto an external wire. Obviously, many possible coupling paths can occur. It is most important that the tool being evaluated can fully model each piece of the problem, the source, the radiating element, and the coupling mechanism.

10.3 Standard Modeling Problems

Although there are many EMI/EMC modeling problems, most of them can be broken down into specific classes. Six different modeling problems are presented here as representative of EMI/EMC modeling in general. While there are many different possible modeling problems, these problems are challenging for most, if not all, vendor tools. They allow the EMI/EMC engineer to clearly discuss the types of problems needing modeling without using standard waveguide or similar problems.

10.3.1 Radiation Through Apertures

One of the most basic EMI problems is to predict the level of attenuation obtained accurately, when using an EMI shield. Far-field equations are frequently used as a starting point for this type of analysis. Since the shield is usually very close to the EM energy source, far-field assumptions are not valid. The interaction of the near fields generated by the source and its conductors with the shield are very dependent on their geometries and need to be understood. Two problem geometries are shown in Figures 10.1 and 10.2. Figure 10.1 shows the most simple

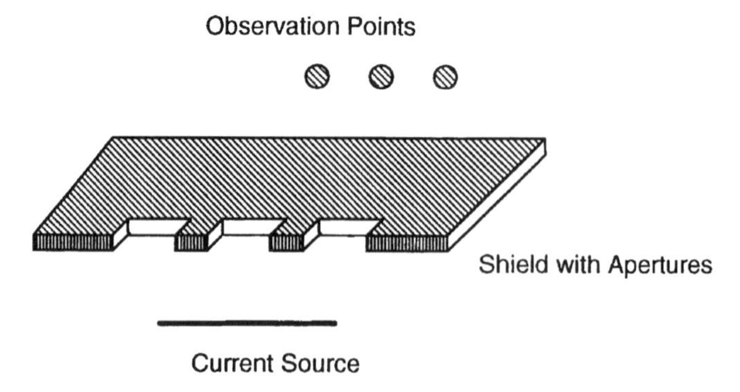

Figure 10.1 Shield With Apertures and Current Source

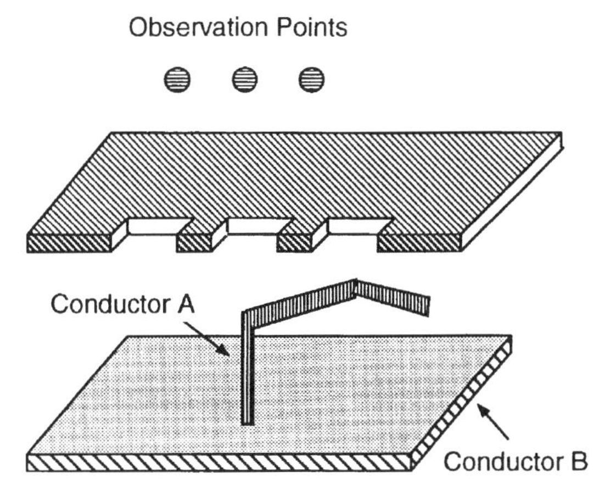

Figure 10.2 Shield With Apertures and Complex Source

case where the source is a current on a straight wire. Figure 10.2 shows a more complex source that comprises two conductors; the excitation for this problem is a voltage between the two conductors.

While the observation points are shown close to the shield, it is important to obtain both near-field and far-field predictions. This enables the data to be used both on the bench and at a radiated test facility. The goal of this model is to predict the field strength both at the observation points shown and some far-field points and to demonstrate how the emission levels are affected by the source conductors.

10.3.2 Wire Through an Aperture

The case of a wire traveling through an aperture, as in the case of a connector through a shielded box, is an important model for problems concerned with common mode voltages on a cable. The geometry for this problem is shown in Figure 10.3.

As shown in Figure 10.3, an external cable is excited from an internal source through the electromagnetic fields to its associated interface connector. There is no direct electrical connection between the internal source and the connector. The connector is shorted (electrically) to the shielded box wall, to make it representative of a perfect capacitive filter. Figure 10.3 is not drawn to scale; it is important to note that the connector loop should be small (10 mm per side) while the external cable should be long (1 m). The goal of this problem is to determine

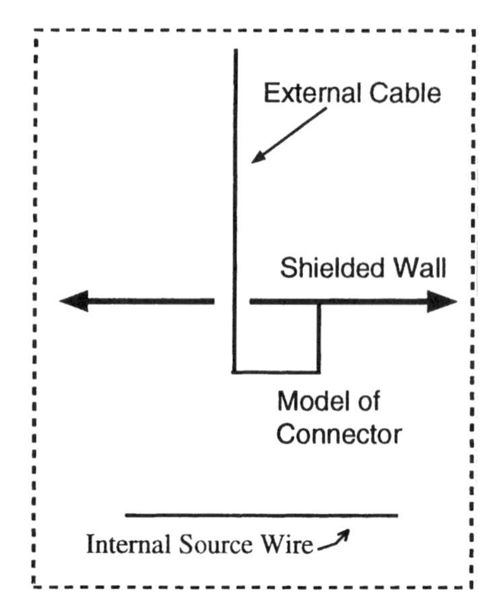

Figure 10.3 Coupling Through an Aperture

the field strength at a distant (10-m) receiving antenna, based on the current in the internal source conductor.

10.3.3 RF Current on Reference Plane Due to Remote Source

Often the use of high-speed (high edge rate) microstrip (or stripline) bus lines creates unacceptably high RF currents on another remote part of the circuit module reference plane. It is desirable to predict these currents, as they can be coupled through connectors or other enclosure openings and result in significant radiated field strength. These currents are sometimes controlled by creating voids on the reference (ground) plane between the area where the high frequency signals exist, and the area where the high frequency signals are not desired. This requires a complete current distribution solution. Figure 10.4 shows the geometry for this problem—an illustrative case for the evaluation of RF current in a remote critical area of a circuit module.

As can be seen in Figure 10.4, the microstrip bus is in an area of the board remote to the critical connector area. Although most of the RF currents will be coupled to the reference plane directly below the microstrip lines, some amount of RF current will exist over the entire

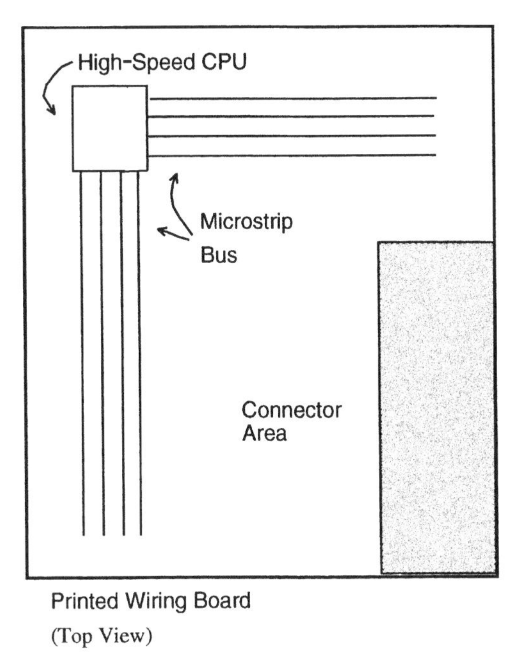

Figure 10.4 RF Current on a Reference Plane

board, possibly causing coupling to the outside through the connector. The goal of this model is to predict both the current distribution and the peak amplitude in the I/O area relative to the microstrip source current.

10.3.4 Common Mode Voltage on a Connector Due to a Known Noise Source

Coupling between a daughter card or microstrip bus to a connector in a shielded box can be accomplished either through conducted RF currents (as in the previous case) or by direct EM fields. The modeling of the common mode voltage present on the connector must take into account the shielded box, the impedance between the circuit module reference plane and the shielded box, and the coupling. The geometry for this case is shown in Figure 10.5. Note the lack of a direct electrical connection between the noise source and the connector.

The goal of this model is to determine the common mode voltage between the connector and the shielded box. This can be used to

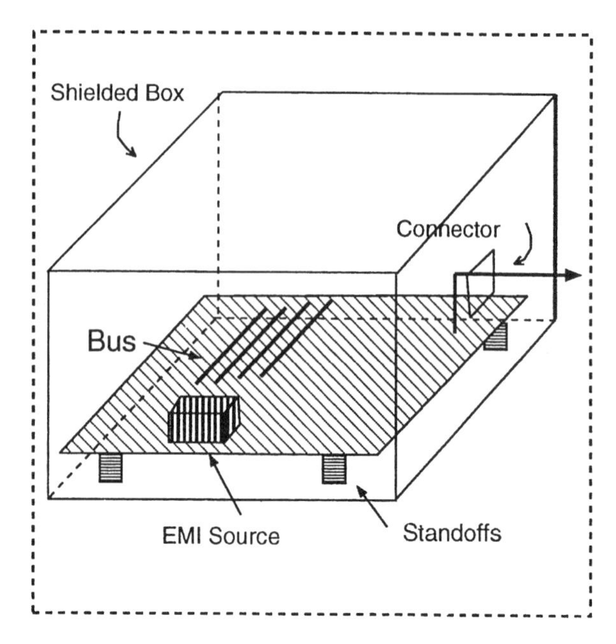

Figure 10.5 Partial Shield Model

predict the final radiated field strength either directly or by the use of a second tool.

10.3.5 Reduction in Coupling Due to Partial Internal Shield

The coupling within a shielded box between the EMI source and an area sensitive to these EMI signals is of concern in many designs. The sensitivity may be due to susceptibility or may be due to a set of I/O cards with connector/cables providing an uncontrollable escape path of the EMI signals. Figure 10.6 shows the geometry for this case.

As can be seen in Figure 10.6, a partial shield is used to reduce the amount of signal strength coupled to the sensitive region with the I/O cards and connectors. This partial shield may be a special piece of metal, or it might often be another circuit module with a solid reference plane with low impedance to the mother board. The goal of this model is to determine the change in field strength due to the presence of the partial shield. It is important to note that there will be points in the frequency response where coupling is enhanced rather than reduced. These are important to detect and should be clearly seen.

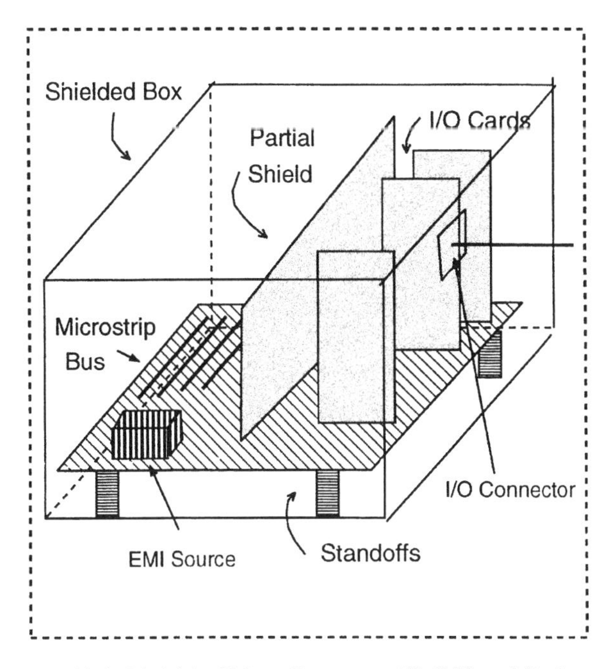

Figure 10.6 Model for Voltage Between an I/O Cable and Enclosure

10.3.6 Direct Radiation from an Unshielded Circuit Module

Not all electronic devices require shielding in order to comply with the regulatory limits, and major cost savings can result from the elimination of unneeded materials. To analyze these situations, it is necessary to have a model that can predict the radiation that comes directly from a circuit module. There are many details that must be addressed for this case. While emphasis is often placed on the signal routing and general module layout, there are other major factors in how much energy will be radiated. These include the finite size and imperfections of the reference plane, the physical size of connector, and components installed over it. The goal of such a model is to ensure that a tool can properly address the imperfections in the ground plane and the true placement of etch. This problem is shown in Figure 10.7.

It is important to include sufficient detail in the circuit module to specify the problem fully. Fortunately, if a module does not require shielding it is probably a relatively low-speed case, and its dimensions will be small compared with the frequencies of concern.

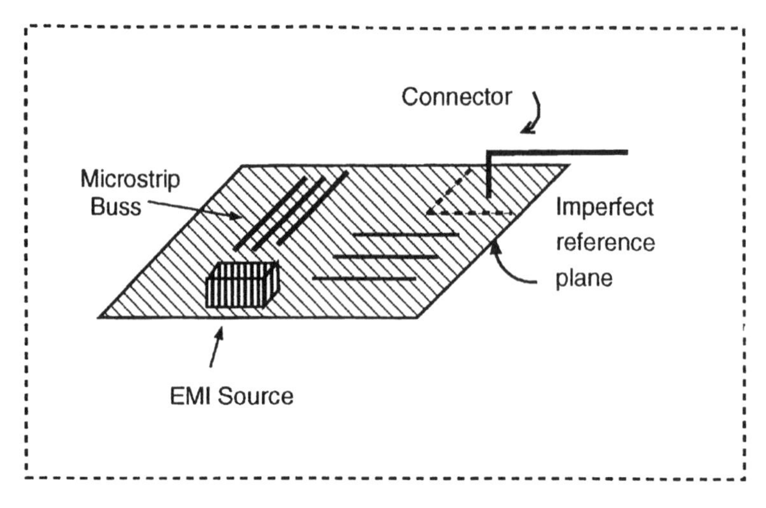

Figure 10.7 Model for Radiation from a Circuit Module

10.4 Summary

A number of cases are provided to be used as standard EMI/EMC modeling problems that can be used to evaluate existing and future vendor software modeling tools. Naturally, not every possible problem geometry can be considered here. However, the standard cases proposed represent a wide variety of possible problems and can be considered representative of the overall range of EMI modeling.

Index